THE ECOLOGY OF THE WYE

MONOGRAPHIAE BIOLOGICAE

VOLUME 50

Editor
J. ILLIES†
Schlitz, F. R. G.

Dr W. Junk Publishers   The Hague Boston London   1982

The ecology of the Wye

R. W. EDWARDS AND M. P. BROOKER

Dr W. Junk Publishers   The Hague-Boston-London   1982

Distributors:

for the United States and Canada

Kluwer Boston, Inc.
190 Old Derby Street
Hingham, MA 02043
USA

for all other countries

Kluwer Academic Publishers Group
Distribution Center
P.O. Box 322
3300 AH Dordrecht
The Netherlands

Library of Congress Cataloging in Publication Data

Edwards, R. W.
  The ecology of the Wye.

  (Monographiae biologicae ; v. 50)
  Includes index.
  1. Stream ecology--Wye, River (Wales and England)
2. Wye, River (Wales and England) I. Brooker, M. P.
II. Title. III. Series.
QP1.P37 vol. 50 [QH144]   574s   82-6580
                      [574.5'26323'094295]   AACR2

ISBN-13:978-94-009-7988-8   e-ISBN-13:978-94-009-7986-4
DOI: 10.1007/978-94-009-7986-4

Cover design: Max Velthuijs

To our parents, whose personal sacrifice
made possible our careers in science

Craig Goch Reservoir in the Elan Valley.

Preface

The valley of the River Wye has long been famed for its natural beauty and the salmon stocks which the river supports. As one of the largest substantially unpolluted rivers in Southern Britain, the Wye, which rises in Wales and flows into England, is now of considerable significance in the strategy for conservation of freshwater habitats in the United Kingdom and is designated as a Site of Special Scientific Interest by the Nature Conservancy Council (N.C.C.). However, despite this scientific importance, published studies of the aquatic ecology of the river during the earlier decades of this century were principally restricted to the excellent series of articles and books on the salmon written by J. A. Hutton during the period 1912–47 and scattered records of the plants of the river banks. During this period the Fishery Board and successive organisations responsible for fisheries, also published reports describing general water quality and salmon catch statistics. In 1970 a survey was undertaken by the N.C.C. which provided the basis for the decision to give it special protected status.

The Wye has been exploited as a major water source for Birmingham, through the development of reservoirs in the Elan Valley, since 1904. Further development of the capacity of the Elan catchment occurred in 1952 with the completion of the Claerwen Reservoir. Proposals to enlarge one of the already existing direct supply reservoirs, Craig Goch, to provide storage to regulate the natural flow of the River Wye drew further attention to the need to know much more about the river's general ecology and water quality. In 1974 the Welsh Water Authority, on behalf of the Craig Goch Joint Committee and Central Water Planning Unit, contracted the University of Wales Institute of Science and Technology (U.W.I.S.T.) to undertake studies of the fish populations, general ecology and water quality of the R. Wye with a view to providing a firm scientific baseline upon which the effects of future proposed changes in flow could be assessed. Other work was also commissioned at this time by the N.C.C., the Water Authority and others, and studies of the aquatic and riparian vegetation (U.W.I.S.T.), algae (University College, Cardiff), birds (Royal Society for the Protection of Birds) and otters (Society for Promotion of Nature Conservation) of the R. Wye have recently been completed. Such wide ranging and yet detailed and contemporaneous investigations of a large river system, like the Wye, have not been previously undertaken in the United Kingdom, and this opportunity has been taken to draw together the results of these investigations in this book in an effort to provide as complete a record as possible of the general ecology of the river and to speculate on the consequences of further changes of land-use and water resources of the catchment.

Much of the work described above is already published in specialist journals and reports, and this book seeks to provide a summary for the general scientific reader: those requiring more detailed description are referred to the original publications which are listed in the bibliography.

Where groups of species all have common names, such as the birds and fish, these are used in the text, but in other cases, such as the flowering-plants, only some have such widely accepted common names and here, for the sake of consistency within the Chapter or Section, the alternative Latin name is adopted.

Apart from the scientists who took part in the studies described in this book, most of whom are cited in references, the authors wish to record their thanks to those many people who were no less important in 'putting and keeping the wagon on the road', particularly the Llysdinam Trustees, General Sir Thomas Pearson and the many land and fishery owners who provided access to the river, Dr. M. A. Learner, Dr. F. M. Slater, Dr. R. Williams and secretarial and technical staff of the Applied Biology Department, UWIST, Mr. M. Owens and Mr. E. M. Staite of the Welsh Water Authority, Dr. J. Hellawell of the Severn-Trent Water Authority, Dr. R. Abel of the Central Water Planning Unit, Mr. J. Beaver of Sir William Halcrow and Partners and Mr. R. Lovegrove of the Royal Society for the Protection of Birds.

Contents

List of plates

1. Geology, land-use and hydrology

Geology

The River Wye (Nant Gwy) rises at 677 m O.D. on Plynlimon in the Cambrian Mountains, close to the source of the River Severn, and drains a total catchment area of about 4180 km^2 before joining the Severn Estuary at Chepstow, about 250 km from source (Fig. 1). Geologically the catchment is characterised principally by impermeable Lower Palaeozoic rocks (mudstones and shales) and the more permeable Upper Palaeozoic rocks (marls and sandstones) which form about 41% and 55% of the total catchment area respectively. The Carboniferous (limestones) and Triassic rocks, which are generally restricted to the south-eastern part of the catchment, make up about 3% and 1% respectively of the total area (Fig. 1).

The junction between the Lower and Upper Palaeozoic rocks forms the major geological division in the catchment, which is reflected generally by a change in the water quality of the R. Wye (see Chapter 2) and probably represents the approximate boundary between the location of a deep, sea-water trough (geosyncline) in the north-west of the catchment and a shallow sea or land in the south-east during Ordovician times. The impermeable Ordovician and Silurian sediments in the north-west of the catchment consist of a great thickness – the Silurian deposits exceed 1000 m near Plynlimon – of shales and conglomerates thrown into a long anticline with a north-east to south-west axis and pitching towards the north-east. This deformation, the Towy anticline, which probably resulted from the Caledonian mountain-building movements, has brought Ordovician rocks to the surface through the overlying Silurian formations (Fig. 1). Vulcanicity, associated with these movements, resulted in moderate deposits of tuffs and dolerite east of Builth Wells which have produced crags resistant to weathering.

Indeed on a local scale the many igneous intrusions in the Builth and surrounding areas have produced mineral-rich springs, a feature of the district which has provided a tourist attraction since the early seventeenth century. In the eighteenth century Llandrindod Wells was a principal resort and visitors from England and elsewhere came to take the waters, the thirty springs, according to guide books of the time, being 'saline, sulphurous, or chalybeate, cold and non-aerated, tonic and laxative'.

On a broader scale the calcareous deposits of the Wenlock beds of the Silurian

1

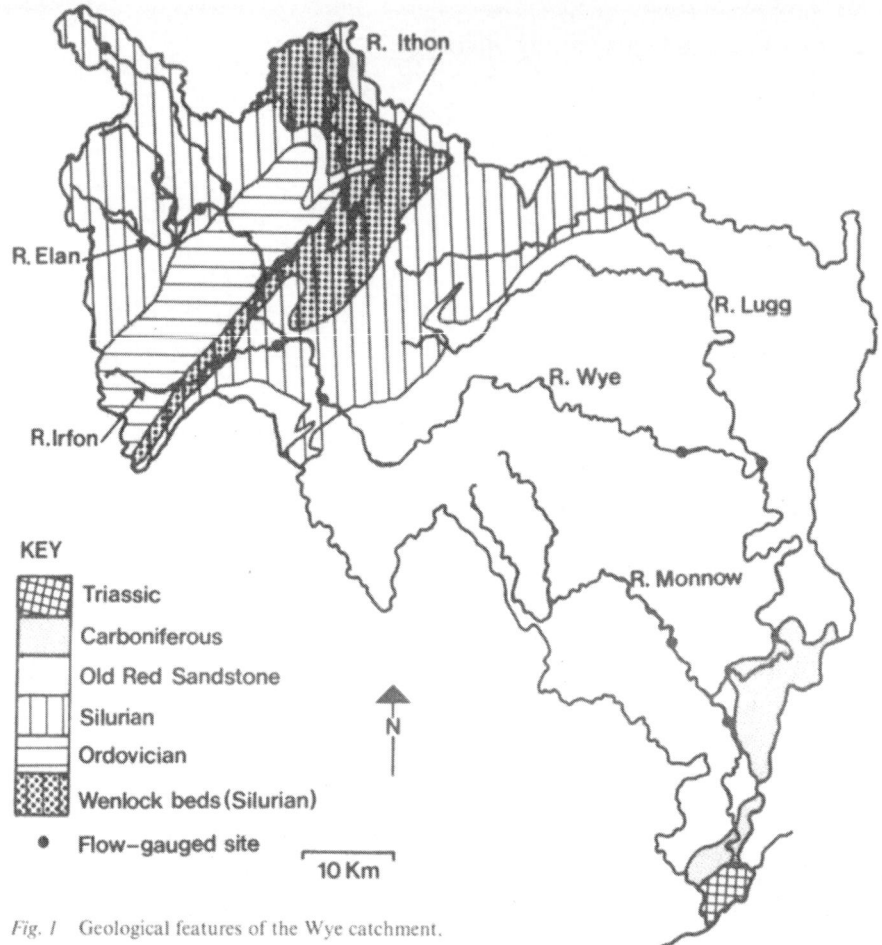

Fig. I   Geological features of the Wye catchment.

series, which lie in a belt from Newtown to Builth Wells (Fig. 1) and overstep the Ordovician rocks in a narrow outcrop in the south-west (George 1970; Wye River Authority 1972), assume some importance in the water quality of the R. Wye (see Chapter 2) and appear to be of some ecological significance too (see Chapter 4).

The rocks of the upper Wye catchment are of particular significance, since they form the basis of many geological classifications. It was in this area that Sir Roderick Murchison established his Silurian system (Murchison 1839) and some of the excitement of his discoveries is captured in his own words:

> Travelling from Brecon to Builth by the Herefordshire road ... low terrace-shaped ridges of grey rock ... appeared on the opposite bank of the Wye and seemed to rise out conformably from beneath the Old Red of Herefordshire. Boating across the river at Cavansham Ferry I rushed up to these ridges and to my inexpressible joy found them replete with transition fossils, afterwards identified with those at Ludlow.

2

Other classical work has been undertaken on the Ordovician igneous and associated rocks of the Builth/Llandrindod inlier (Jones 1946, 1948), the most notable find being that of an Ordovician shoreline (Jones 1949).

The south-east of the Wye catchment is characterised by underlying rocks of the Old Red Sandstone (ORS) series which consist of three groups: two in the Lower and one in the Upper ORS. The older group of beds in the Lower ORS consists predominantly of siltstones and sandstones with some thin limestones. Generally this older group of rocks is fairly easily weathered and forms low-lying plains, such as those around Hereford (see Fig. 2), but where sandstones occur their greater resistance to erosion results in higher ground. The second group in the Lower ORS is the Breconian which is found principally in the south-west of the catchment and is composed chiefly of sandstones which are resistant to weathering, resulting in higher ground. The small areas of Upper ORS are found in the lower Wye valley and these rocks, mainly Tintern Sandstone with a basal quartz conglomerate, are also hard-wearing and give rise to steep cliffs in the Wye Gorge (Welsh & Trotter 1960)

The small areas characterised by Carboniferous rocks are in the south-east part of the Wye catchment and consist of calcareous shales and mudstones overlain by thin limestones. Later deposits of Drybrook Sandstone were in turn overlain by the Upper Carboniferous coal-bearing strata consisting of shales, sandstones, grits and coal seams (Wye River Authority 1972).

The Triassic rocks of the Wye catchment represent only about 1% of the total area and occur as small outcrops of conglomerates and marl (Welsh & Trotter 1960) at the mouth of the river (Fig. 1). Major ice movements during the glacial period of the Pleistocene probably originated from a centre of glaciation in the north-west of the catchment around Plynlimon and spread south-east on to the low-lying plains around Hereford (see Fig. 2). Other contributions of ice probably came from North Wales and the Lake District. Glacial deposits are thickest and most continuous in the northern part of the catchment and, in places, boulder clay up to 12 m deep effectively masks the pre-glacial topography. Glacial sands and gravels also cover considerable areas, particularly on the lower ground, and are highly permeable to water, contributing considerably to the base flow of streams in certain areas (Wye River Authority 1972).

Relief and drainage pattern

Most of the features of the area do not result from the underlying geological structure but from the varying resistance of rock types to the effects of weathering on a landscape uplifted in Tertiary times. For example, the peak of Plynlimon probably represents an exposure on resistant Ordovician rocks in the core of a structural dome rising above the remnants of a series of plateaux typical of Mid-Wales: the High Plateau (520 610 m O.D.), the Middle Peneplain (365–490 m

3

O.D.) and the Low Peneplain (245-305 m O.D.) are probably a succession of marine surfaces resulting from Pleistocene sea level changes (Newson 1976). It was Sir Andrew Ramsay, climbing up out of a tributary valley near the source of the Wye and Severn, who first recognised the significance of these upland plateaux during the late nineteenth century.

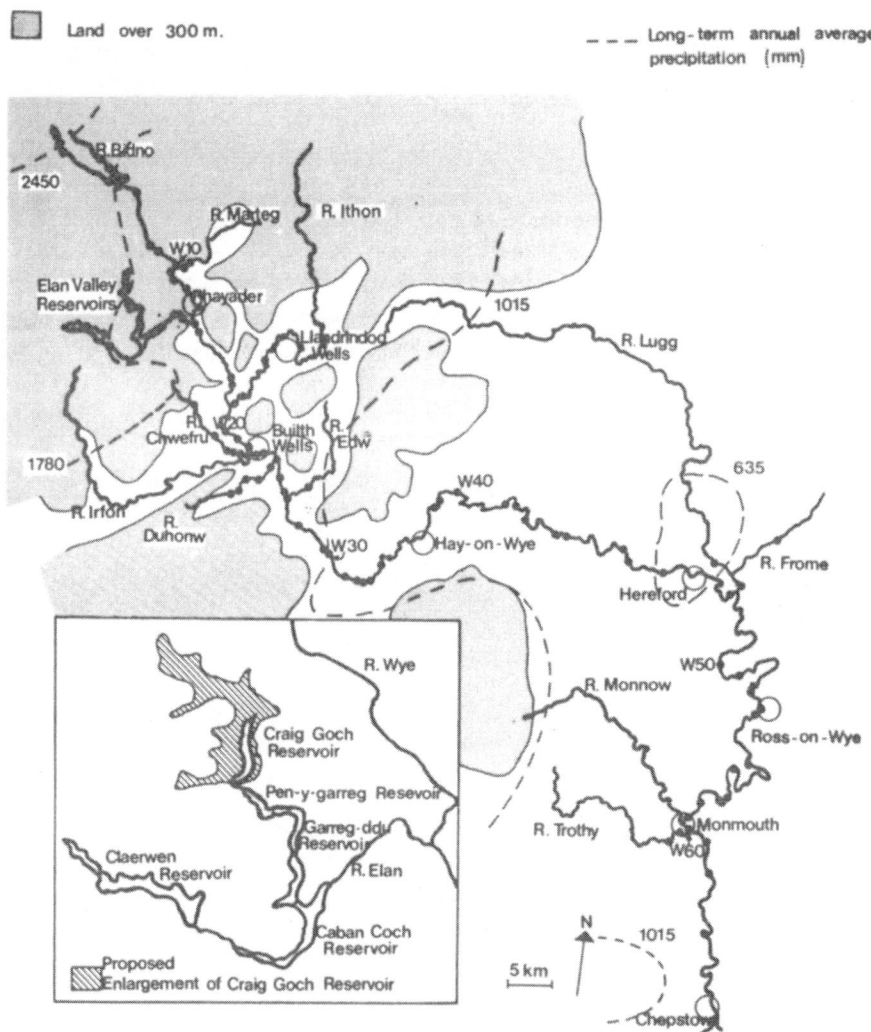

Land over 300 m.

— — — Long-term annual average precipitation (mm)

Fig. 2 Sampling sites (●) for chemical, plant, invertebrate and fish surveys of the R. Wye and its tributaries (in the text Ir = Irfon, E = Elan, L = Lugg), and altitude and rain distribution within the catchment. Insert shows the existing reservoirs of the Elan Valley together with the extent of the proposed enlarged Craig Goch Reservoir.

Generally, the uplands, above about 300 m and consisting of Silurian and Ordovician rocks, occur in the north-western part of the catchment (Fig. 2) and include the Plynlimon range (up to 753 m). Further south the plateau of the Black Mountains, composed of ORS (predominantly sandstones resistant to weathering), forms a scarp overlooking the Wye Valley rising to a height of 811 m. The more easily weathered ORS forms the Hereford Plain, much of which is below 70 m O.D.

From its source the R. Wye drains first an area of blanket peat and then flows through upland grassland before being confined to a steep V-shaped channel, characterised by sequences of pools and riffles and fringed by trees, until it reaches Builth Wells, 60 km from source. Most of the tributaries in the uppermost reaches are swift-flowing torrents in young, steep-sided, V-shaped valleys with relatively large bed gradients but the three major tributaries which join the Wye in this section show a variety of characteristics. The River Elan is impounded – most of its reservoirs being constructed during the period 1895–1904 to provide water for Birmingham and the remaining reservoir, Claerwen, being built between 1946 and 1952 to supplement the water resources of that city. The Ithon is a left bank tributary with a catchment area of 365 km^2 whose lower elevation and subdued relief contrasts with the upper Wye, and the Irfon is a right-bank tributary of rather smaller catchment area (244 km^2) with high relief.

South-east of Builth Wells the much enlarged Wye enters a gorge section and runs over bedrock. Subsequently the valley floor widens and the channel, which now has a shallower gradient (see Fig. 11), is characterised by steep alluvial banks with nearby ox-bow lakes representing previous meanders. Below Hereford the Wye receives drainage from its largest tributary, the River Lugg, which has a catchment (1070 km^2) of generally subdued relief, broad valleys and low altitude.

Below its confluence with the R. Lugg the Wye meanders considerably even after entering a number of gorge sections downstream of Ross-on-Wye where, although entrenched to a depth of about 180 m, the river maintains a meander course unrelated to the underlying geological structure (Miller 1935; Welsh & Trotter 1960). These gorges, which are often densely wooded, are best known at Symonds Yat and the sharpness of the gorge here suggests a relatively recent origin (Welsh & Trotter 1960).

In Monmouth the Wye receives the Monnow, which is its second largest tributary with respect to catchment area (433 km^2) but not to flow (Table 1). Part of the Monnow catchment lies in the upland area of the Black Mountains and the remainder consists of low-lying relief similar to the valleys of the lower Lugg and its tributaries.

It is clear that in the upper reaches of the Wye and elsewhere the tributaries have suffered considerable capture and Brown (1960) concluded that the R. Wye may have received at one time the waters of the upper Severn and the R. Clywedog, now in the Severn catchment. Further downstream it was thought that the R. Lugg originally joined the River Teme, a tributary of the R. Severn, until the Ice Age

5

Table 1 Flows at selected gauging stations in the Wye catchment. Data from Wye River Authority (1972). The locations of flow gauging stations are shown in Fig. 1

Flow gauging station	Distance from source of the Way	Drainage area to gauging station (km²)	Average flow (m³/s)	Average flow per unit area (m³/s/km²)
River Wye				
Pant Mawr (W1)	7	27	1.8	0.061
Ddol Farm (W14)	36	174	7.0	0.040
Erwood (W28)	68	1280	38.3	0.030
Redbrook (W61)	220	4010	71.6	0.018
River Lugg				
Lugwardine	-	886	10.6	0.012
River Irfon				
Cilmery	-	244	9.9	0.041
River Ithon				
Disserth	-	358	7.5	0.021
River Monnow				
Kentchurch		357	5.8	0.016

when they were separated by a glacier (Wye River Authority 1972). It seems that the present course of the Wye, particularly the incised meandering south of Hereford, is different from that in pre-glacial times, but the history is too complex to separate the effects of glaciation from earlier influences (Earp & Hains 1971).

Precipitation and river flow

There is a strong correlation between relief and annual average rainfall, which is the principal form of precipitation (Newson 1979), in the Wye catchment (Fig. 2). The main rain-bearing winds originate in the south-west and, since elevation in the catchment generally decreases from west to east, this results in higher rainfall in the west and lower rainfall in the east. Thus, the 1015 mm isohyet and 305 m contour coincide and include the Black Mountains and the upland areas to the north and west of the Wye Gorge at Erwood (Fig. 2). In consequence, the catchment experiences considerable spatial extremes of annual rainfall ranging from above 2500 mm on Plynlimon to less than 660 mm in the vicinity at Hereford, locations separated by a distance of less than 80 km. Such differences in rainfall are reflected in the water resources of catchments which lie adjacent to each other. For example, the Irfon catchment has an average annual rainfall of over 1660 mm, whereas the Ithon catchment receives less than 1100 mm (Wye River Authority 1972).

The high rainfall and impermeable nature of the geology in the upper reaches of the River Wye result in high rates of run-off per unit of catchment area and rapid changes in flow, e.g. at Pant Mawr (W4) and Ddol Farm (W14) (Table 1 and Fig. 3). Further downstream, changes in the rate of precipitation and drainage geology alter the flow characteristics in the lower catchment and rates of run-off and change of flow are considerably lower. On average, over half of the total catchment run-off originates in the upper third of the catchment.

At their confluences with the main river the major tributaries, the Ithon, Irfon and Lugg, contribute about 50, 40 and 20% respectively to the total flow of the River Wye below each confluence. The Elan catchment, impounded by a series of direct supply reservoirs (Fig. 2), exports about 50% of its annual run-off to Birmingham via a gravity-fed aqueduct. The reservoirs in the Elan Valley generally provide a compensation flow of 1.34 m^3/s to the river downstream of the dams and subsequently to the R. Wye, but severe drought conditions during 1975 and 1976 led to reductions in such compensation flows, down to 0.52 m^3/s at certain periods. Of course, flows in the River Elan are substantially greater at times of impoundment overspill, generally during the winter period (Fig. 3).

The drought conditions pertaining in the River Wye, and more widely in the U.K. in 1975 and 1976 have been considered fully elsewhere (Wright 1976; Wye River Division 1977). The 18-month period February 1975 – September 1976 was generally the driest recorded in the Wye catchment (Wye River Division 1977) with

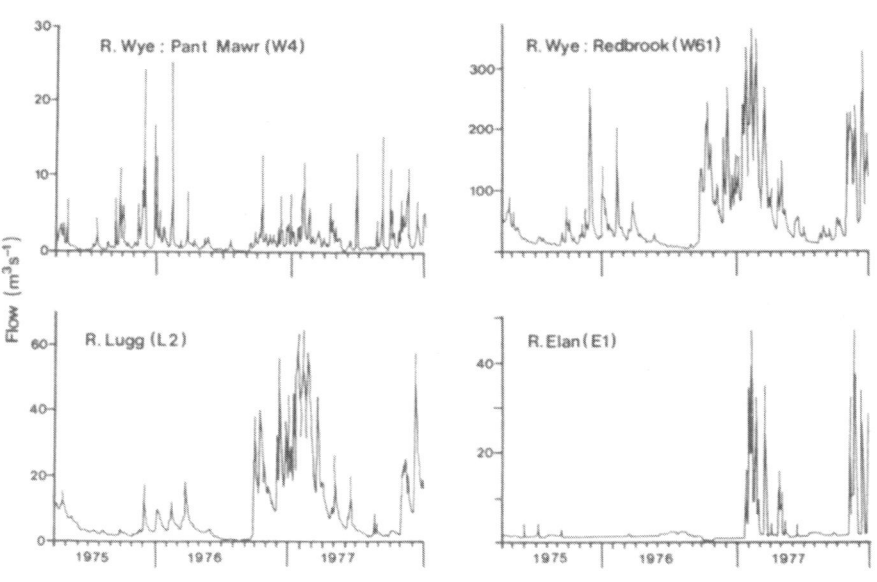

Fig. 3   River flows at four recording stations in the Wye catchment during the period 1975–77.

the most severe conditions during the three-month period June – August 1976, particularly in the southern and eastern parts of the catchment where river flows were the lowest ever recorded. Analysis of average rainfall for the 18-month period of drought yielded a return period in excess of 1000 years.

Records from Redbrook flow gauging station, 220 km from source, indicated that annual run-off in 1975, 1976 and 1977 was 70, 71 and 110% of the long-term average (1933–1974). However, natural flows at the same site in the 12-month period, September 1935 – August 1976, constituted only 38% of the long-term average. The driest period throughout the catchment was in August 1976 when natural flows at Redbrook represented about 9% of the long-term average for that month. During the period of severe drought the flow of the River Wye was regulated from the Caban Coch Reservoir in the Elan Valley, and reservoir releases during the lowest natural flows contributed 52% of the total flow at Redbrook in the lower reaches of the R. Wye.

Population and land-use

The Wye catchment is sparsely populated and the total population was estimated to be 196 000 (0.47/ha) in the 1971 census. The major centres of population are Hereford (46 500) and Chepstow (8000), and only four other towns have populations exceeding 3000 – Leominster, Ross-on-Wye, Llandrindod Wells and Monmouth. These urban centres contain about 40% of the total population of the catchment. Of the current estimated total population of about 200 000, about 60% are connected to sewers.

Agriculture is the main industry in the Wye catchment, and its form is related to the relief, rainfall and underlying geology of the region. Overall, based upon 1970 data, the proportional distribution of land-use in terms of the total catchment area was:

	%
Tillage	20
Temporary grass	14
Permanent grass	37
Rough grazing	15
Woods and forests	9
Urban land	3

Data extracted from maps, based on the Second Land Utilisation Survey, by subcatchments C1–C9 (Table 2) indicated that the high rainfall areas of the north-west with high relief are principally used for rough grazing contrasting with the nearby, heavily afforested (42%) upper Severn catchment. The lowland areas of the Wye catchment, with richer soils derived from the weathered ORS rocks, support mixed and dairy farming and horticulture.

Table 2 Categories of land-use in the Wye sub.catchments expressed as a percentage of the total land of the sub-catchment. Data obtained from Second Land Utilisation Survey. Catchments C1 to C9 are shown in Table 6 (Chapter 2)

Catch-ment	Heath	Grass	Wood-land	Arable	Market gardening	Urban
C1	66.0	2.0	32.0	..	–	–
C2	61.5	25.6	10.1	2.7	–	–
C3	89.0	3.7	6.9	0.4	–	–
C4	26.5	58.5	9.8	5.5	–	–
C5	38.9	30.2	30.2	1.5	–	–
C6	30.0	55.6	8.4	3.6	–	1.6
C7	12.9	65.7	9.5	10.9	0.6	0.4
C8	7.9	54.7	9.9	17.1	6.8	2.6
C9	4.1	58.5	11.7	18.3	5.0	2.5
Total catchment	19.0	52.5	11.4	12.0	3.2	2.0

The R. Wye is already exploited directly as a source of water both for agricultural irrigation (2000 $\times$ 10^3 m^3 in 1971), which is obviously chiefly restricted to periods of low river flow, and for potable supply. The major abstractions for the latter are at Llyswen (about 2 $\times$ 10^3 m^3/d), Byford (about 2 $\times$ 10^3 m^3/d), Hereford (about 30 $\times$ 10^3 m^3/d), Lydbrook (about 36 $\times$ 10^3 m^3/d) and Monmouth (about 3 $\times$ 10^3 m^3/d). The substantial precipitation in the north-west of the catchment has been exploited by reservoiring the valleys of the River Elan and River Claerwen (Fig. 2). The three reservoirs in the Elan Valley – Caban Coch (including Garreg Ddu), Pen-y-Garreg and Craig Goch, completed at the beginning of this century, provide a total storage capacity of about 50 800 $\times$ 10^3 m^3 (Table 3). Claerwen Reservoir, opened in 1952, provides an additional 48 300 $\times$ 10^3 m^3 of storage.

Table 3 Characteristics of the reservoirs of the Elan and Claerwen valleys

Reservoir	Dam height (m)	Top water area (ha)	Capacity (m^3 $\times$ 10^3)
Elan			
Caban Coch	37.2	202	35530
Pen-y-Garreg	37.5	50	6050
Craig Goch	36.6	88	9230
(Proposed enlarged			
Craig Goch	96.0	796	245000
Claerwen			
Claerwen	56.1	263	48300

About 320×10^3 m^3 of water are carried daily by gravity through an aqueduct, which is about 120 km long, to Birmingham for direct supply. Raw water from the reservoirs receives some preliminary treatment (liming and filtration) before distribution and backwash water from the filters, after flocculation with ferric sulphate and settlement, is discharged to the R. Elan.

The potential of using the unexploited water resources, which are not already reservoired, has been recognised for some time. For example, in 1894 a scheme was proposed, but not approved for construction, in which waters of the upper Wye, Ithon, Yrfon (Irfon) and Edw were to be transported by aqueduct to the lower Thames Valley for utilisation by the then Metropolitan Water Board of London. More recently plans to regulate the flow of the River Wye in order to support abstractions at Monmouth in the lower reaches of the river have been made. These, at present tentative, proposals, together with their possible ecological consequences, are discussed in Chapter 7.

2. Water quality

Introduction

Some of the first investigations of water quality in the Wye catchment were of the spa waters in Mid-Wales, mineral-rich waters draining from igneous intrusions in the Ordovician and Silurian sediments (see Chapter 1). These spa waters, widely regarded as having therapeutic properties, created a holiday industry in the area during the nineteenth century, with principal centres at Llanwrtyd Wells, Builth Wells and Llandrindod Wells. The spa facilities remain at Llandrindod and efforts have been made recently to revive interest in the use of such waters. Analyses of spring waters from Llandrindod in a guide of 1906 (Anon 1906) are shown in Table 4. The flow of such springs represents a very small proportion of the flow of rivers within the area, and their effects on the water quality are very local.

On a broader scale the pristine waters of the upper Wye and its tributaries have been the focus of attention of those responsible for water supplies since the mid-nineteenth century, the earliest major water resource developments being manifested in the existing Elan Valley reservoirs (see Chapter 1).

The waters of this predominantly rural catchment are, of course, free from major sources of sewage and industrial pollution and until relatively recently, when proposals to regulate the flow of the R. Wye and abstract considerable amounts from the lower reaches were considered, there seemed little need to undertake a comprehensive surveillance programme of river water quality. Consequently, except for a few simply-measured parameters such as ammonia, nitrate, ortho-phosphate, alkalinity and BOD, few measurements were regularly and extensively made.

The basic description of water quality in the R. Wye is derived principally from surveys during the period 1975-1977 (Edwards et al. 1978; Oborne et al. 1980) supplemented by data, collected since 1970 by the Welsh Water Authority, from a site at Monmouth. In addition, special ancillary studies have been undertaken in the upland tributary, R. Elan (Paull 1978), to describe the behaviour of iron and manganese in relation to discharges from the Elan reservoirs and a water treatment works, and in the lowland tributaries, R. Frome and R. Trothy (Houston & Brooker 1980), to determine nutrient fluxes in small catchments with productive farmland and comparatively intensive urban settlement and high population

11

densities. The local effects of the impoundments in the Elan Valley on water temperatures have also been described.

General water quality of the River Wye

The water quality of the R. Wye is largely related to soil structure and the underlying geology of the catchment – which determines the potential source of most materials (except sulphate and sodium, much of which are atmospheric in origin, and phosphate which is derived from sewage) – and to the pattern of rainfall, infiltration and run-off which determines the leaching and dilution of such materials. The nature of land-use influences not only the amount of run-off – for example, the upland pastures yield about 15% more run-off, with similar precipitation, than the coniferous plantations of the neighbouring Severn catchment – but it also modifies water quality. Chemical inputs are further influenced by discharges of sewage effluent, although in the Wye catchment with a total population of only 200 000 and only about 60% being connected to sewers, the effect of such sewage and sewage effluent is small except with respect to phosphate concentrations.

Chemical quality may also be modified in situ by chemical processes, such as calcium precipitation at high pH, and biological processes, such as silica utilisation by diatoms and denitrification. Such changes are displayed in temporal variations in water chemistry.

Table 4 Analyses of three well waters at Llandrindod Wells in the late nineteenth century (concentrations in mg/l: T = trace)

Constituent	Spring		
	Saline	Sulphur	Chalybeate
Sodium chloride	4762	2320	3965
Potassium chloride	29.9	T	17.1
Calcium chloride	975	701	922
Magnesium chloride	37.1	37.1	195
Lithium chloride	T	T	T
Calcium carbonate	49.9	128.3	8.6
Iron carbonate		–	18.5
Ammonium carbonate	2.9	–	1.4
Calcium nitrate	5.7	10.0	8.6
Calcium sulphate	17.1	7.1	10.0
Ammonium sulphate		2.9	–
Potassium bromide	0.3	2.9	T
Potassium iodide		T	T
Silica	24.2	34.2	18.5
Iron & aluminium oxides & phosphates	17.1	1.4	–

Spatial distribution. As in many rivers, concentrations of most determinands increase steadily downstream in the R. Wye (Fig. 4 and Table 5). The water draining the mineral-poor impermeable mudstones in the upper catchment have low concentrations of minerals and nutrients and the lower catchment, particularly that part draining the arable lands of the Lugg basin, is characterised by much richer water. Generally, increases in concentrations are greatest at sites downstream of Hereford. Thus, NO_3-N and PO_4-P show twelve-fold and sixteen-fold differences in concentration respectively between the most upstream and downstream sites sampled (Fig. 2). Mean calcium concentrations increase from about 2 mg/l 7 km downstream from the source to about 30 mg/l near the estuary.

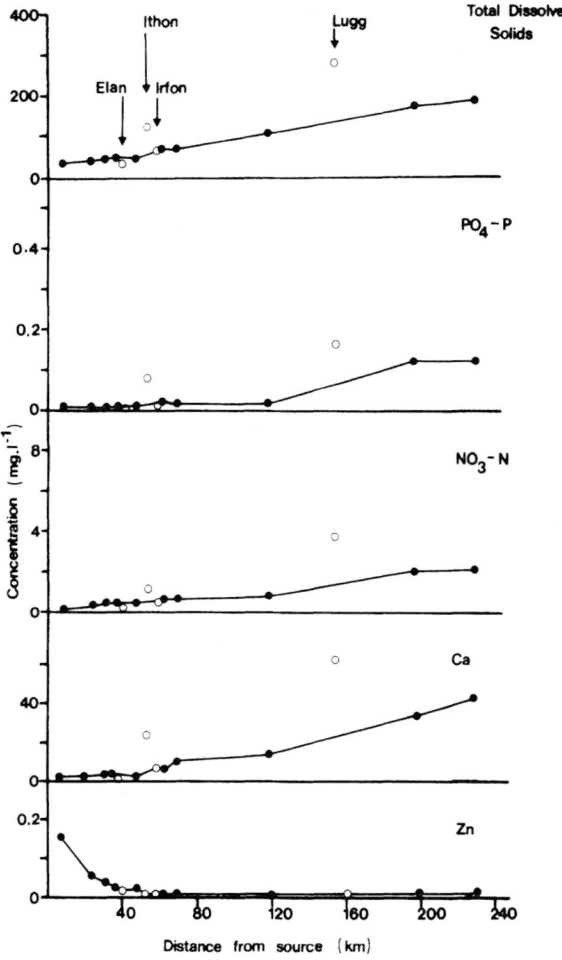

Fig. 4 Average concentrations of total dissolved solids, orthophosphate-phosphorus, nitrate-nitrogen, calcium and total zinc at sites along the R. Wye and major tributaries.

Table 5 Mean values of water quality determinands at selected sites calculated for a low-flow year (1975–76), D) and a high-flow year (1976–77, W) (all concentrations in mg/l)

Deter- minand	Site W4		Site W14		Site W18		Site W12		Site W28		Site W61		Site L2	
Ca	2.42	1.60	3.90	3.92	3.20	3.40	20.0	18.5	9.53	10.1	30.4	35.6	54.1	54.9
Na	5.01	4.31	7.14	6.61	5.25	5.62	10.1	8.33	6.22	6.32	11.9	9.76	13.3	11.6
K	0.56	0.17	0.88	0.49	0.67	0.43	2.55	1.67	1.04	0.80	2.50	2.18	3.28	2.88
Mg	1.11	1.02	1.48	1.56	1.36	1.52	5.03	4.60	2.35	2.53	5.40	5.96	7.91	7.14
PO_4-P	0.010	0.002	0.016	0.004	0.011	0.005	0.085	0.098	0.024	0.011	0.153	0.090	0.204	0.100
NO_3-N	0.10	0.13	0.52	0.65	0.37	0.61	0.93	1.27	0.52	0.91	1.21	5.24	2.90	4.62
SiO_2	2.66	1.82	2.84	2.01	2.40	1.86	3.03	2.31	2.41	2.07	2.65	2.71	5.13	3.87
SS	0.94	1.76	1.36	9.28	1.61	7.69	8.60	27.5	4.00	20.3	7.99	88.6	4.40	27.9
SOC	1.44	1.80	1.46	2.13	1.76	2.23	2.95	3.14	2.21	2.50	2.53	2.82	2.12	2.56
TDS	38.4	35.6	51.3	55.0	45.3	49.6	132.0	121.0	74.4	78.6	175.0	200.0	282.0	281.0

SS = Suspended solids; SOC = Soluble organic carbon; TDS = Total dissolved solids.

14

The major upland east-bank tributary, the Ithon, is ionically richer than the upper Wye, reflecting drainage from the calcareous Wenlock beds (Fig. 1). However, the upland west-bank tributary, the Elan, is generally less rich than the Wye and the quality of the main river below the Elan-Wye confluence is locally influenced by the acid, nutrient-poor water discharged from the impoundments of this tributary, particularly during periods of low natural flow when the proportion of Elan compensation water is high.

In contrast to most determinands, concentrations of heavy metals, such as zinc and nickel, are lower at downstream sites, the zinc concentration, for example, falling from about 0.150 mg/l at 7 km from source to about 0.025 mg/l at 30 km from source (Fig. 4). Small streams near to the river source and draining old lead mine workings are major contributors of these metals.

One aspect of particular ecological significance relates to the relatively high concentrations of iron and manganese in the R. Elan, the mean concentrations of 0.5 Fe and 0.1 Mn mg/l being about ten times greater than those in the upper Wye. Paull (1978) estimated that about 75% of the total iron and 50% of the total manganese in the R. Elan were associated with particulate matter, and almost half of the load of these metals is derived from a water treatment works and not directly from the reservoirs. Deposition rates equivalent to 0.26 Fe and 0.16 Mn g/m^2/d were estimated from concentration changes within the river, but measured rates of accretion from sediment deposition experiments were appreciably lower. These deposits assume considerable significance in affecting the distribution of macroinvertebrates in the R. Elan (see Chapter 4).

Temporal variation. The water chemistry of the R. Wye is characterised by marked seasonal changes, particularly in the lower catchment where concentrations are substantially higher than the upper catchment (Table 5). For example, alkalinity and the concentrations of total dissolved solids and the major ions, such as calcium, magnesium, sodium, potassium and bicarbonate are generally lowest during the winter and highest during the summer. These differences are predominantly related to the seasonal pattern of flows, concentrations being diluted at high river flows. Simple concentration-fhow regressions of these major ions, after logarithmic transformation, generally account for more than 40% of the variation (Oborne et al. 1980) (Fig. 5).

Different input sources result in different temporal behaviour of some nutrients. For example, throughout the catchment highest concentrations of phosphorus occur principally during summer low-flow periods (Fig. 6), and concentrations were particularly high during the drought of 1976. Since the major source of phosphorus is relatively constant, being derived from sewage effluents, changes in concentration within the river principally reflect the dilution which such effluents receive.

This behaviour contrasts with that of nitrate which generally has highest concentrations during winter high-flow periods (Fig. 6). Such an increase in concentration

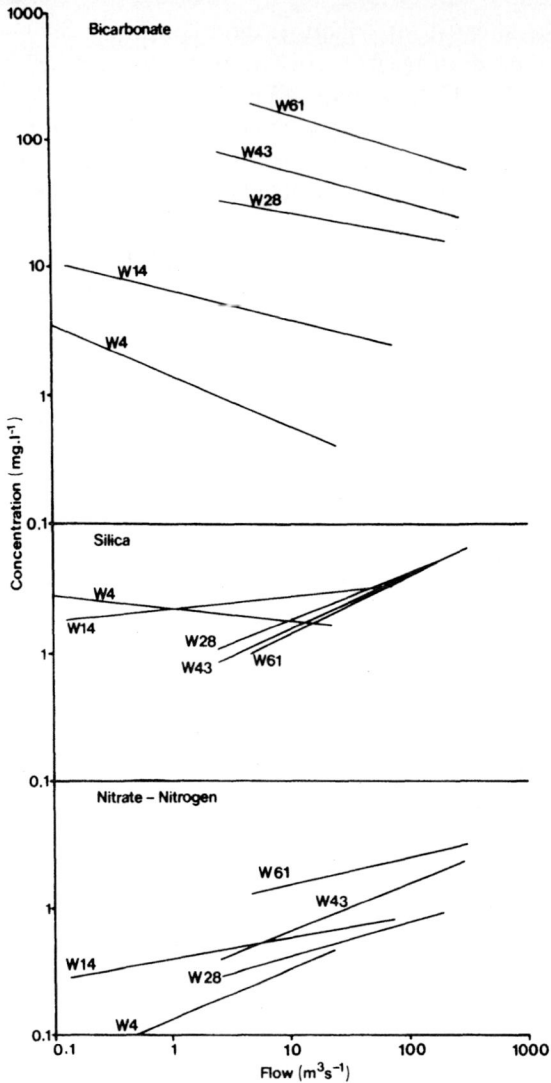

Fig. 5 Concentration flow relationships for bicarbonate, silica and nitrate-nitrogen at sites on the R. Wye (see Fig. 2).

was particularly marked during the autumn of 1976 when a period of drought was terminated by heavy rainfall. Substantial increases in nitrate concentration (0.1 to 5.0 mg/l) were recorded in some lowland reaches: these probably resulted from flushing processes following the prolonged mineralisation of organic material in the soil during the long dry summer. However, Houston and Brooker (1980) found that during 1978 nitrate concentrations in one lowland tributary, the Frome, increased during the low-flow summer period, and it was concluded that elevated

16

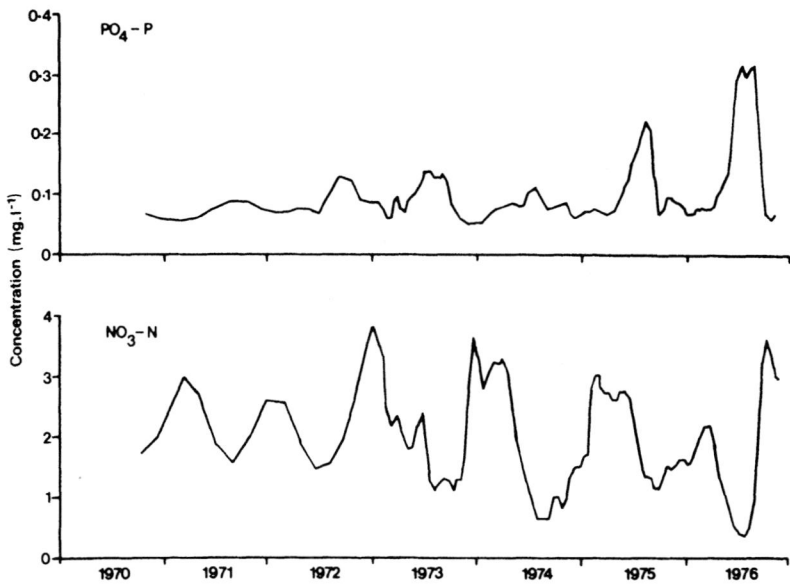

Fig. 6 Seasonal variation in concentrations of orthophosphate-phosphorus and nitrate-nitrogen at Monmouth.

summer concentrations were derived from the higher proportion of nitrate-rich ground water which, at this time of the year, received little dilution from surface run-off (Fig. 7).

Some seasonal changes in water chemistry may be directly attributed to biological activity which in turn may be related to flow. For example, concentrations of silica are always highest during winter with major depletions generally observed

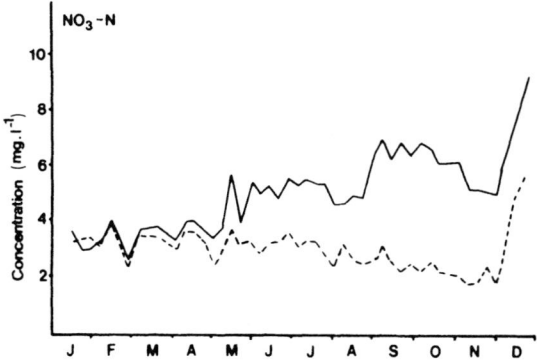

Fig. 7 Seasonal changes in nitrate-nitrogen concentrations in the Rivers Frome (——) and Trothy (----), 1978.

17

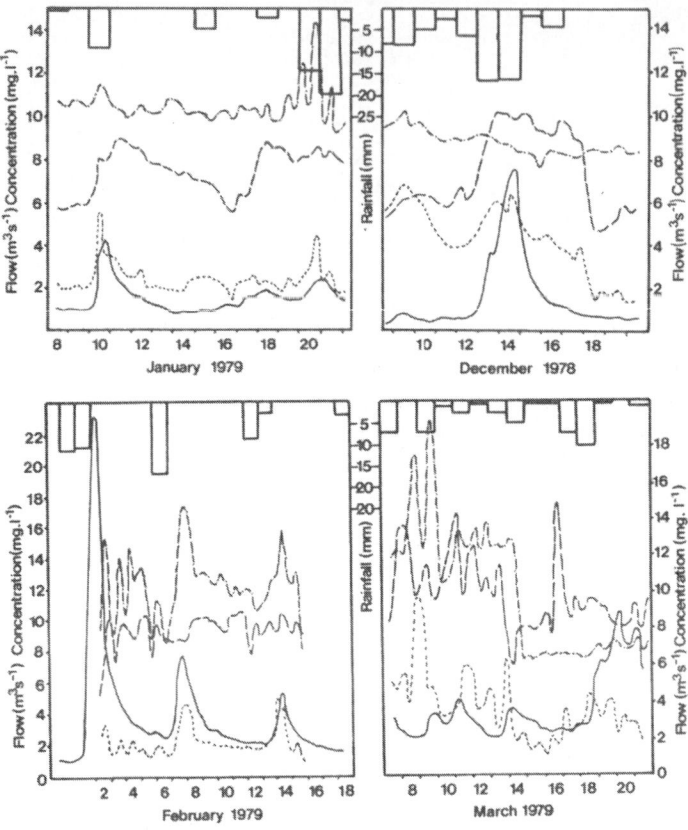

Fig. 8 Changes in the concentration of nitrate-nitrogen (---), orthophosphate-phosphorus $\times$ 10^{-1} (......) and silica (- · · · ·) in response to changes in flow (——) in the River Frome. The histograms indicate the daily rainfall (mm).

in spring and late summer, probably reflecting patterns of diatom growth. Similar behaviour has been observed in rivers in eastern England (Edwards 1973a; Toms et al 1975) and from the lowland tributaries of the Wye, the Frome and Trothy (Houston & Brooker 1980).

As water quality is related to flow as well as to other characteristics which vary seasonally, it is not surprising that there are appreciable differences between wet and dry years. Table 5 shows the mean concentrations of several determinands for contrasting 12-month periods of wet and dry conditions. In the case of some determinands (Ca, Na, K, Mg) major differences between the two periods occur in the headwaters and in the case of others (NO$_3$, SS) in lowland reaches: phosphate and silica concentrations are very different throughout the length of the river.

Intensive sampling programmes (8 h intervals) on the R. Frome have revealed

18

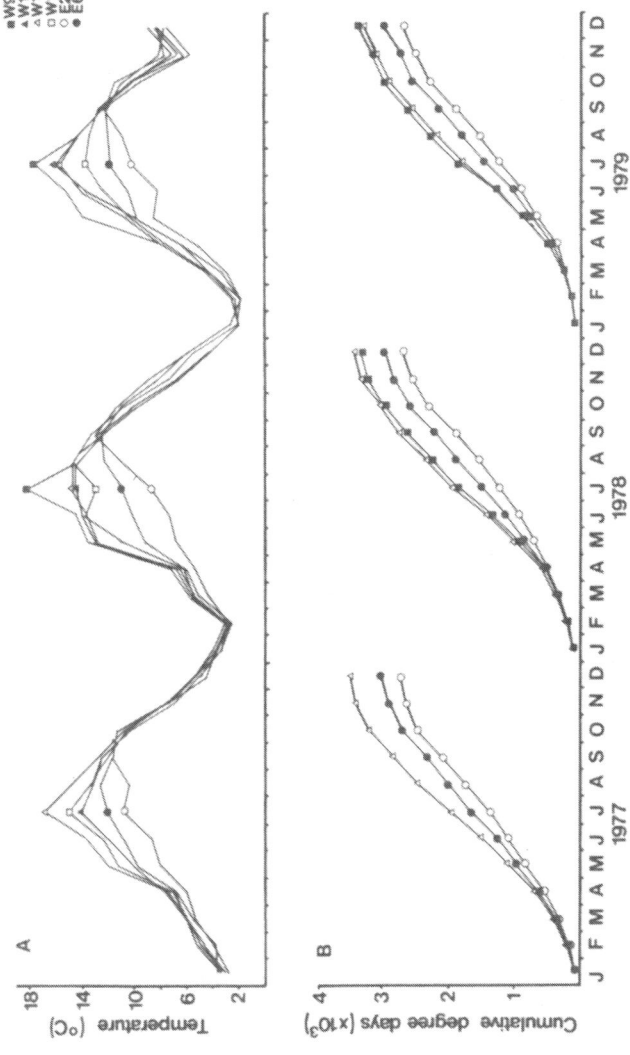

Fig. 9 (A) Mean monthly temperatures at sites in the R. Elan (E2, E6) and in the R. Wye upstream (W9, W15) and downstream (W17, W19) of the confluence (for location see Fig. 2) in the period 1977–79. Data from W9 not available in 1977. (B) Cumulative degree-days at selected sites.

short-term changes in water quality and whilst some of these seem clearly related to storm events (Fig. 8), such as the increases in nitrate concentration at the periods of increased flow in December 1978 and January 1979, others are not. Furthermore, responses to flow increases are by no means consistent as is shown in Fig. 8 by the relative constancy of nitrate concentrations during storm events in February 1979. The meteorological history is crucial in understanding the behaviour of those substances, such as nitrate, derived from pedological processes.

The impoundments on the R. Elan influence not only the water chemistry at downstream sites (see above) but, locally change the temperature regime of the river. Mean annual water temperatures (1977–79) at two sites on the R. Elan, at the dam and about 7 km downstream, were 1.7 and 1.0 °C colder respectively than the average temperature (9.1 °C) at a site on the Wye immediately upstream of the Elan–Wye confluence. About 3 km downstream of the confluence mean temperatures were reduced by about 0.4 °C, but no differences were detected at a site 5 km further downstream.

Mean monthly temperatures at all sites reached lowest values (2–3 °C) in January and February and increased at similar rates until April after which site differences were marked (Fig. 9). At Wye sites, particularly above the Elan confluence, water temperature increased rapidly and reached a peak of 14–19 °C in July whilst in the Elan mean temperatures increased at a slower rate, and the lower peak values (11–13 °C) were delayed until September. The greatest differences in the mean monthly temperatures (6–7 °C) between sites occurred during May–July, the period of major temperature increase, and the time of annual maximum temperatures in the Elan was delayed by about 70–110 days when compared with the unimpounded upper reaches of the R. Wye. Similar local effects have been reported for rivers below impoundments in the north of England (Lavis & Smith 1972; Crisp 1977).

Catchment losses. Oborne et al. (1980) estimated catchment losses of several determinands during two 12-month periods of contrasting hydrology – a dry period from September 1975 to August 1976 (D) and a much wetter period from September 1976 to August 1977 (W). Nine flow-gauged sites in the catchment, which were sampled for water quality, conveniently divided the catchment into sub-catchments (C1–C9, Table 6), appropriate subtractions being made where flow-gauged sites represented more than one sub-catchment.

There are substantial differences in yield between different parts of the catchments, related to geology and land-use, and between the wet and dry periods (Table 7). Annual yields per unit catchment of several substances, e.g. potassium, calcium, magnesium, nitrate and phosphate were greater in the lower Wye and Lugg than elsewhere, and this difference was generally most marked during the wet period. In contrast, highest yields of sodium (80 kg/ha/y), probably derived from rainfall, were in the upper reaches of the river which received the highest annual precipitation.

20

Table 6 Sub-catchments of the Wye catchment (see Fig. 2)

Catch-ment	Area (km²)	Population density 1[a]	Population density 2[b]	Description
C1	27.2	0.09	0.005	source to site W4
C2	146.8	0.13	0.009	site W4 to W14
C3	184	0.09	0.006	Elan catchment
C4	358	0.25	0.13	Ithon catchment
C5	244	0.14	0.06	Irfon catchment
C6	320	0.15	0.06	catchments to site W28 (C1 + C2 + C3 + C4 + C5)
C7	620	0.22	0.11	site W28 to W43
C8	1,030	0.48	0.23	Lugg catchment
C9	1,080	1.07	0.78	site W53 to W61
Total catchment	4,010	0.50	0.31	source to site W61

[a] Population density 1 = total population density (persons/ha)
[b] Population density 2 = sewered population density

A separate comparison of losses of phosphate and nitrogen from the Frome and Trothy catchments during 1978 confirmed the large yields of these nutrients from the lowland east bank catchments like the Frome (Houston & Brooker 1980). In particular the greater urban development and higher population in the Frome catchment were reflected in the phosphate losses (0.47 kg/ha/y) compared with the Trothy (0.22 kg/ha/y). In the Wye the yield of phosphate could be accounted for solely from sewage effluents whilst less than 20% of nitrate was derived from these sources, even during the wet period (Oborne et al. 1980). Table 8 compares yields from other studies with those from the Wye catchment.

The daily per capita outputs of phosphate and nitrate can be calculated from the established relationships between sub-catchment yields and population density. These were 1.8 (dry) and 3.1 (wet) g PO_4-P and 6.8 (dry) g NO_3-N, generally similar to values estimated from other studies (Smith 1976).

Water quality model of the R. Wye. Whilst in some sub-catchments such as the Ithon, an upland tributary, relatively high proportions of the variance of some determinands can be explained in terms of flow (dissolved solids, 67%; magnesium, 81%; bicarbonate, 92%) in other sub-catchments the proportions are considerably smaller (Oborne et al. 1980). Therefore, the use of simple concentration-flow regressions are inadequate as predictive models of water quality. Alternative techniques, based upon the separation of river flows into baseflow, run-off and effluent components were first used to model the behaviour of a number of determinands in the Rivers Severn and Avon, explaining up to 93% of the variation in concentration in some cases (Birtles 1977; Birtles & Brown 1978).

21

Table 7 Sub-catchment yields (kg/ha/yr) and runoff (Ml/ha/yr) for the periods *D* and *W*

	Period	C1	C2	C3	C3*	C4	C5	C6	C7	C8	C9	Total catchment
TDS	*D*	536.7	217.2	86.4	308.4	337.2	495.7	380.3	327.4	327.4	538.9	383.8
	W	597.6	351.7	166.5	388.5	797.9	702.9	964.4	867.7	1347.0	1244.6	1030.9
NO_3-N	*D*	2.77	4.07	0.39	1.92	3.45	3.98	2.78	3.94	3.92	5.49	4.05
	*E**	0.1	0.9	2.1	0.4	7.3	2.9	4.4	9.1	1.4	36.9	17.7
	W	2.80	6.45	0.44	1.97	9.93	6.14	9.93	16.30	27.20	18.00	16.70
	*E**	0.1	0.6	1.8	0.4	2.5	1.9	1.2	2.2	1.1	11.2	4.3
PO_4-P	*D*	0.066	0.036	0.013	0.076	0.085	0.054	0.021	0.057	0.161	0.529	0.207
	*E**	1.0	28.5	17.0	2.9	68.3	60.9	131.9	216.0	52.7	139.4	120.5
	W	0.076	0.060	0.024	0.087	0.156	0.083	0.086	0.230	0.538	0.894	0.444
	*E**	1.0	17.2	9.2	2.5	37.0	39.7	31.5	53.8	15.8	82.5	56.2
SiO_2	*D*	28.8	15.5	4.5	15.1	9.4	21.9	5.1	8.0	5.0	-2.9	5.3
	W	33.8	25.3	8.8	19.4	32.1	33.7	30.6	8.3	35.7	14.5	23.3
SS	*D*	20.3	16.2	4.4	24.5	29.0	29.9	32.3	25.0	7.87	41.5	25.0
	W	21.4	26.0	8.0	28.1	90.7	45.8	94.2	114.6	92.1	424.9	175.7
Na	*D*	70.0	37.9	10.7	52.4	23.6	40.9	23.4	11.3	14.4	27.2	21.6
	W	80.1	61.0	19.5	61.1	55.6	62.4	142.7	2.7	56.8	52.3	52.9
K	*D*	4.4	4.0	0.8	2.9	4.6	4.5	4.1	4.1	3.6	7.4	4.8
	W	4.9	6.0	1.3	3.4	10.6	6.8	10.8	10.4	15.8	15.2	12.3
Ca	*D*	27.4	25.8	4.1	16.6	50.2	48.9	45.0	71.7	62.9	91.6	64.3
	W	32.0	40.5	4.8	17.3	117.2	73.5	93.9	156.4	276.7	130.1	154.6
Mg	*D*	15.1	10.3	2.6	13.0	11.8	18.5	8.9	5.9	8.9	19.0	11.8
	W	17.5	16.5	3.7	14.1	26.7	28.4	27.8	17.7	35.6	43.8	30.9
HCO_3	*D*	20.3	32.8	1.99	–	128.9	113.4	139.3	232.5	224.4	393.1	230.4
	W	23.9	48.5	4.30	–	266.1	166.3	273.8	426.0	780.0	712.4	516.0
Runoff	*D*	15.70	5.84	2.70	9.64	3.03	7.62	2.91	0.93	1.19	1.08	2.15
	W	17.60	9.95	5.77	12.71	7.81	11.70	9.41	4.87	5.47	6.12	6.72

C3* = Yield of C3 including yield to supply; *E** = Percentage of catchment yield from sewage effluent.

Table 8 Comparison of Wye catchment yields (kg/ha/yr) from period W with yields from other catchments

Source	River	Na	Ca	Mg	K	SiO$_2$	PO$_4$-P	NO$_3$-N
This study	Wye (C1)	80.1	32.0	17.5	4.9	33.8	0.08	2.8
	Lugg (C8)	56.8	276.7	35.6	15.8	35.7	0.54	27.2
	Wye total catchment	52.9	154.6	30.9	12.3	23.3	0.44	16.7
Edwards (1973 b)	Yare	73.0	302.0	12.0	9.0	10.0	0.40	14.0
	Tud	44.0	204.0	8.0	6.0	7.0	0.20	10.0
Smith (1976)	Main	95.7	132.1	66.5	11.7	88.9	0.65	13.7
Hughes & Edwards (1977)	Cynon	231.0	98.7	67.4	27.0			
Sutcliffe & Carrick (1973 a)	Duddon	76.5	74.0		5.2			
Crisp (1966)	Rough Sike	45.2	9.0		3.1			
Troake et al. (1976)	Slapton Wood							26.7
	Stokley Barton							29.4
	Slapton Stream							27.0
	Gara							33.6
Owens et al. (1972)	Gt. Ouse							15.6
	Trent							26.8

23

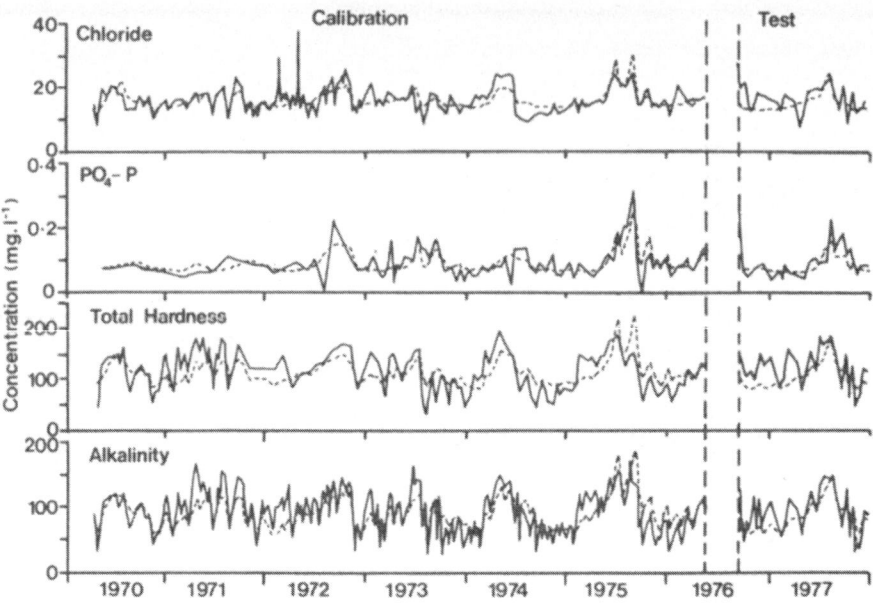

Fig. 10 Performance of the water quality model (----) compared with the collected data (——) during the calibration (1970–76) and test (1976–77) periods.

Using a similar approach, Oborne (1981) successfully modelled the behaviour of chloride, phosphorus, alkalinity and hardness in the R. Wye at Monmouth. River chemistry and flow were measured directly, the base-flow component being estimated from the natural hydrograph, and total effluent returns were calculated from licensed consent flows which were assumed to be constant. Mass flows of each flow component were used, after correction for the effect of storage in the river channel, to estimate the natural concentration. The model, based upon data collected during 1970–76, explained between 59 and 88% of the variance in the test data for the period 1976–77 (Fig. 10), and the performance was considerably better than that of flow-concentration regressions calculated from the same data. The major source of error is the separation of flow into baseflow and run-off, and it is significant that the model of PO_4-P, principally derived from sewage, and less susceptible to errors than the other flow components, explained the highest proportion of the variance.

Water quality of the Elan Valley reservoirs

The Elan reservoirs have not been extensively studied. Thompson (1954) examined Birmingham Water Department temperature data for the period 1945–51, and Hopper (1978) made ancillary investigations of temperature and oxygen concentrations at Caban Coch Dam during an examination of zooplankton populations (see Chapter 4). Further data on surface water quality are available from the Severn-Trent Water Authority.

General water quality. Surface water samples collected from Craig Goch and Claerwen reservoirs indicate the acidic nature of the water which drains from moorland areas and generally contains low concentrations of nutrients and minerals and high concentrations of iron and manganese (Table 9). There are few major differences between the general water quality of the reservoirs and that of the River Elan downstream of the impoundments.

Table 9 Mean chemical concentrations (mg/1 except where stated) in surface water samples collected monthly from Claerwen and Craig Goch 1977-79. Data from Severn-Trent Water Authority. Water quality recorded from the R. Elan is shown for comparison

	Claerwen	Craig Goch	Elan
pH (range)	4.8-7.4	4.8-6.9	5.1-6.6
Nitrate-N	< 0.5	< 0.5	0.22
Orthophosphate-P	0.027	0.016	0.009
Silica (SiO_2)	1.24	1.22	1.73
Conductivity ($\mu S/cm$)	39.6	36.3	41.6
Total hardness ($CaCO_3$)	11.3	10.6	8.5
Calcium (1979 only)	1.1	1.0	1.8
Total iron	0.32	0.25	0.50
Total manganese	0.22	0.28	0.10

Vertical stratification. The limited data available indicate that Caban Coch Reservoir, the downstream impoundment in the series which discharges to the R. Elan, stratifies thermally during the period June to September. This can result in temperature differences of up to 11 °C between the water surface and the dam outlets, which are located about 7 m from the bottom of the 37 m dam. The effects of these discharges on downstream river temperatures have been described earlier. The stratification is not stable and some mixing of the waters takes place throughout the summer. Overturn in stormy weather rapidly leads to homogeneous temperatures, and mixing is generally complete by the beginning of October.

Such thermal stratification does lead to depletion of oxygen resources in the deeper parts of the reservoirs, although the minimum oxygen concentration recorded by Hopper (1978) in the summer of 1977 was 5.0 mg/l. Substantially greater oxygen depletion has been recorded in deeper Mid-Wales reservoirs (Llyn Clywedog, 0.2 mg/l oxygen at 65 m), and this is associated with increases in the concentration of iron (>2.0 mg/l) and manganese (>1.0 mg/l). Other reliable quality data are not generally available for the Elan Valley reservoirs, but reducing conditions were recorded in Claerwen Reservoir, the deepest reservoir (dam height, 56 m), resulting in ammonia concentrations of 1.2 mg/l in the compensation flow in August 1951 (Thompson 1954).

25

Plate 1   Making net collections in the Wye Gorge near Erwood (W29).

26

3. Plants

Plant distributions within the river

There were several botanists in the nineteenth and early twentieth centuries who described the aquatic and bank-side flora of limited sections of the Wye. Armitage (1914), Bryan (1894), Gissing (1853) and Monnington (1889) listed species present in the lower reaches near Ross-on-Wye and in the Carboniferous Limestone gorge at Symond's Yat. Monnington recorded the presence of *Ranunculus fluitans* at Ross where it is still abundant (see p. 42), although now regarded as a component of the *R. penicillatus* complex which includes *R. fluitans, aquatilis, peltatus* and their hybrids. Ley, Vicar of Sellack and a very active local botanist who was deeply involved in the production of the Flora of Herefordshire of 1889, described plants in some upper and middle reaches in a series of records between 1874 and 1891. Riddelsdell (1910) concentrated on riparian plants of the locality near Boughrood. Some of those recorded by Riddelsdell (*Serratula tinctoria, Listera ovata, Orobanche major, Euphorbia amygdaloides, Plantanthera bifolia, Dactylorchis maculata* and *Thalictrum minus)* are no longer found in the locality. Ley recorded the presence of the chive (*Allium schoenopraesum*) near Aberedw flourishing 'for a mile or more of distance': more recent studies such as that of Bougourd & Parker (1975) have shown a very marked increase in the local distribution of this species.

Other recent descriptions of plants of the main river have been few and either concerned with a specific site of particular interest (Perring 1957) or based on a cursory survey of aquatic species at relatively few sites (Ratcliffe 1977). However, in 1979 Merry completed a detailed study, commissioned by the Nature Conservancy Council, of both riparian and aquatic vegetation of the main river from source to mouth. Studies of the tributaries include local surveys of aquatic plants undertaken by Haslam (1978) in the R. Lugg and, more systematically and extensively, by Holmes (personal communication) in the R. Monnow, Lugg, Irfon and Elan. The following section is based principally on Merry's study of the main river.

Merry selected 49 sites, each 1 km in length, which, because they were reasonably evenly distributed along the river, are principally at altitudes below 150 m (Fig. 11). Sites were visited to obtain lists of aquatic and riparian species between May and September in 1976 and 1977. The determination of the uppermost limit of

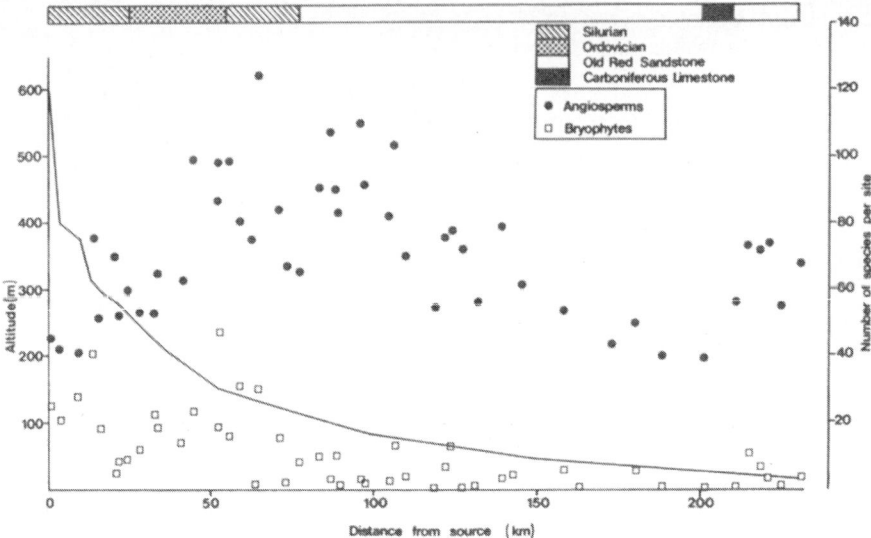

Fig. 11 Numbers of angiosperm and bryophyte species found at sites in the Wye and on its banks in relation to distance from source and altitude. Solid line represents river bed altitude.

banks was difficult at some sites, and in these cases the position of the nearest mature trees or the top of eroded banks fixed the boundary. During further summer visits to each site, the vertical distribution of riparian species was established, species being assigned cover values.

In his surveys of tributaries, Holmes, visiting sites only once during the summer, similarly recorded species within 1 km reaches, but at regular 7 km intervals. In contrast to Merry, Holmes defined his banks as only those areas which were submerged for at least 50% of the year – so reducing his records to aquatic and amphibious species. The resulting contrast in species richness between the two surveys is shown dramatically in Appendices 1 and 2 which list the macrophytes found: Holmes recorded only 60 and 30% as many bryophytes and angiosperms respectively as Merry.

In describing the aquatic and riparian vegetation along a river about 250 km long, it is inevitable that many factors, some interactive, will influence the distribution of species. Additionally, interpretation and explanation are made more difficult when some of the plant communities, such as those on shingle, represent seral stages in a dynamic sequence. Apart from biogeographical considerations, which will impose constraints on the general pool of species available to contribute to the plant associations, climatological, pedological, geological and biotic factors will operate, sorting and selecting species from that pool. These, interacting with anthropogenic factors, will also determine adjacent terrestrial plant communities which, in turn, will influence both local riparian conditions, such as shading, and

28

the local pool of seeds and other propagules. Thus the distributions of some species will be very restricted whilst others will be extensive.

During Merry's survey, 380 species of vascular plants and 114 species of bryophytes were recorded (Appendices 1 and 2): these represent about 18 and 12% respectively of the so-called non-critical* species of these groups in Britain. The vascular plants were taxonomically diverse at the Family level and, although species within the Compositae and Graminae accounted for 20% of the total, the 12 individual families with most species still only accounted for about 57% of all the species recorded.

Sites with the highest richness of vascular plants (>80 species/km) were between 40 and 100 km from the source at an altitude range of 80–180 m: sites richest in bryophytes (>20 species/km) were all within 65 km of the source and above 130 m (Fig. 11). Despite the richness of upstream sites in bryophytes, and mid-stream sites in vascular plants, over 60% of all species occurred within the lower half of the river below an altitude of 90 m, the uppermost 50 km and altitude range 160–610 m adding only 50 species not found elsewhere along the river.

Bryophyta

Unlike angiosperms, bryophytes do not divide easily into those species which may be regarded either as aquatic or terrestrial. Most live under moist conditions, and many might be regarded as amphibious, tolerating long periods of immersion in water without damage. Of the 114 species recorded, the following 44 species are generally most associated with particularly wet stream margins or submerged habitats, and seven species (*) might be regarded as truly aquatic:

Sphagnum auriculatum	*Pohlia delicatula*
Atrichum crispum	*Climacium dendroides*
Polytrichum commune	*Fontinalis antipyretica*
Fissidens crassipes	*Fontinalis squamosa*
Campylopus atrovirens	*Thamnium alopecurvum*
Dichodontium pellucidum	*Acrocladium stramineum*
Dicranella squarrosus	*Brachythecium plumosus*
Cinclidotus fontinaloides	*Brachythecium rivulare*
Rhacomitrium aciculare	*Brachythecium rutabulum*
Rhacomitrium aquaticum	*Drepanocladus exannulatus*
Physocomitrium pyriforme	*Drepanocladus fluitans*
Bryum pseudotriquetrum	*Drepanocladus uncinatus*
Mnium punctatum	*Eurhynchium riparioides*

* Taxa which are readily identifiable by those with botanical training and for which the opinion of an expert, though desirable, is not essential.

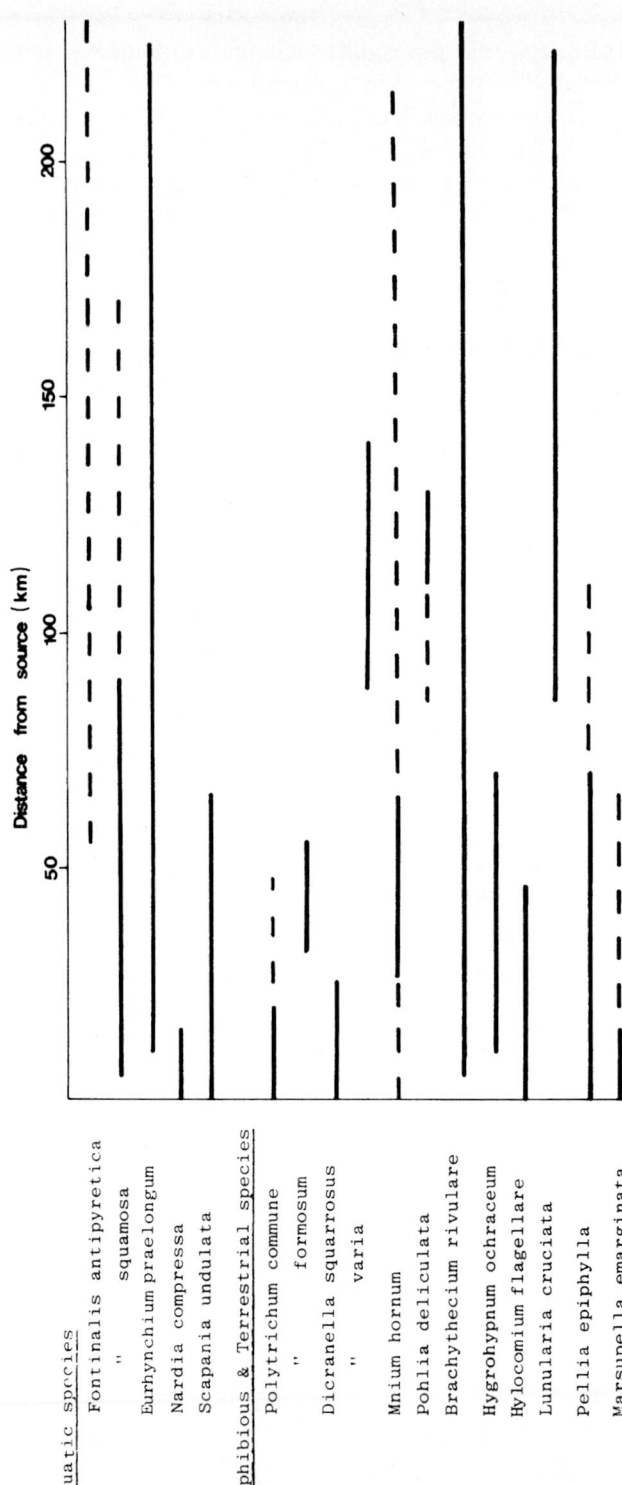

Fig. 12 Distribution of several aquatic, amphibious and terrestrial bryophytes in the Wye and on its banks. Broken line represents zone of patchy distribution.

Philonotus fontana
Pohlia annotina
Hygrohypnum ochraceum
Hylocomium flagellare
Conocephalum conicum
Lunularia cruciata
Marchantia polymorpha
Pellia epiphylla
Riccardia pinguis

*Amblystegium fluviatila
*Amblystegium tenax
Barbilophozia floerkei
Chiloscyphus polyanthos
Diplophyllum albicans
Marsupella emarginata
*Nardia compressa
*Scapania undulata
Solenostoma triste

The distribution of aquatic species, apart from *Amblystegium fluviatile* and *A. tenax* which were only found at single sites, are shown in Fig. 12. *Nardia compressa* and *Scapania undulata* are restricted to higher altitudes, the former only being at high moorland sites. *Fontinalis squamosa* too is principally associated with the upper part of the catchment, unlike *F. antipyretica,* which occurs mainly at lower sites. A selection of amphibious and terrestrial bryophyte species is also shown in Fig. 12: these species may be similarly restricted in their distributions, such as *Polytrichum commune* and *P. formosum, Dicranella squarrosus* and *Pohlia deliculata,* or widespread, such as *Mnium hornum* and *Brachythecium rivulare. Pohlia delicatula,* generally found by streams in wet clay, is interestingly restricted to that region of the Wye on the soft marls of tha Old Red Sandstone series and *Trichostomum sinuosum,* mostly associated with shaded walls in calcareous districts, is restricted to one site on the Carboniferous Limestone of the lower reaches around Symond's Yat.

In other cases species found in riparian habitats are characteristic members of well-defined terrestrial communities which fringe river margins: *Polytrichum formosum,* for example, is characteristic of acid woodlands, particularly sessile oak, and this possibly explains why it is restricted to parts of the upper Wye.

There are some rather unusual discoveries in Merry's survey. For example, *Fissidens crassipes,* generally found in calcareous streams, was recorded on Old Red Sandstone where the average calcium concentration of river water was only about 15 mg/l. *Philonotus calcarea,* also more characteristic of calcareous bogs and wet limestone ledges, was recorded in the upper reaches at a site where *Bryum alpinum, Rhacomitrium aquaticum, R. fasciculare* and *Grimmia doniana* – normally associated with the acid uplands on siliceous rock – were also present.

The bryophyte distributions in the Wye and in tributaries of the catchment, the latter recorded by Holmes, generally conform. Upland species, such as *Nardia compressa, Scapania undulata, Fontinalis squamosa* and *Marsupella emarginata,* were only found in the tributaries at high altitude (Irfon, Elan) and other species more widely distributed in the Wye, such as *Mnium hornum,* were recorded in both upland and lowland tributaries (Appendix 2 and Fig. 12). Nevertheless there were anomalies, such as *Eurhynchium praelongum* and *Brachythecium rivulare,* which were extensively distributed in the Wye but only found in upland and lowland tributaries respectively.

Pteridophyta

The ferns, in the riparian habitat, are generally neither common nor diverse. Merry found only nine species, all widespread in Wales, and of these five were at fewer than three sites. Of these locally rare species in the riparian habitat *Lycopodium selago,* generally regarded as an arctic-alpine species, was only present at the highest site (600 m) and *Phyllitis scolopendrium,* common on limestone substrate, was only found at one site near the Carboniferous Limestone gorge at Symond's Yat. The remaining four widely distributed species, *Equisetum arvense, Pteridium aquilinum, Dryopteris filix-mas* and *Athyrium filix-femina,* are not particularly associated with the riparian habitat except perhaps *A. filix-femina* which prefers moist sites often near water.

Angiospermae

Unlike the bryophytes, the flowering plants may be separated into truly aquatic species 'those which normally stand in water and must grow for at least a part of their life-cycle in water, either completely submersed or emersed' (Muenscher 1944) and those which tolerate or even prefer periodic submergence or wet soil around their roots. There are, of course, those few species which seem ambivalent by this classification, and perhaps the polymorphic *Polygonum amphibium* is good example. This section is divided therefore into two parts with a separate treatment of aquatic and riparian species. The riparian species too may be divided into those normally expected along water-courses or at least near water, e.g. *Caltha palustris* (Marsh marigold), *Viola palustris* (Marsh violet), *Impatiens glandulifera* (Indian balsam), *Narthecium ossifragum* (Bog asphodel), and those adventitious species more commonly associated with other habitats but catholic or adventurous in their behaviour, e.g. *Ranunculus repens* (Creeping buttercup), *Capsella bursa-pastoris* (Shepherd's purse), *Spergula arvensis* (Corn spurrey), *Chrysanthemum vulgare* (Tansy).

It might be expected that the recorded aquatic and truly riparian species would be more widespread, i.e. occur at more sites, than the adventitious riparian species. This does not seem to be the case however (Fig. 13), all categories showing similar distributions with about half the species occurring at fewer than six sites.

It might also be expected that the aquatic species would be more continuously distributed than their riparian neighbours, chemical factors to which both shoots and roots are generally exposed showing few sharp local discontinuities, and seed dispersal, at least in a downstream direction, being aided by water transport. There

32

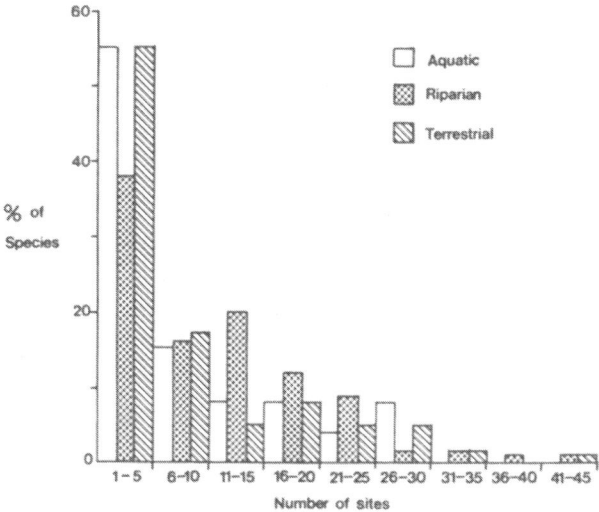

Fig. 13 Frequency distribution of the number of sites at which aquatic, amphibious and terrestrial angiosperms occur.

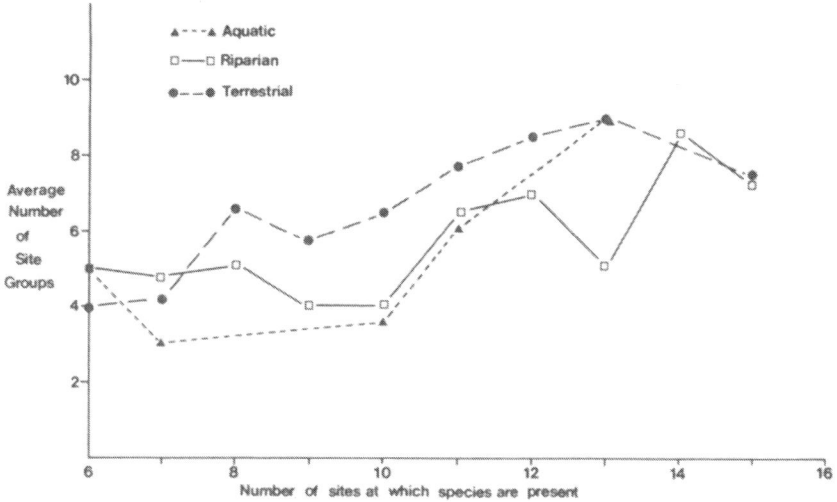

Fig. 14 Average number of groups of contiguous sites (stretches) compared with total number of sites at which terrestrial, riparian and aquatic species occur.

are, however, other local site factors which impose spatial patchiness in aquatic plant distributions, perhaps the most important being water velocity which affects not only the response of shoots, but also of roots through its influence on bed stability and composition. Local patchiness in riparian and adjacent terrestrial communities confers further aquatic heterogeneity by shading, which affects both stream temperature and photosynthetically active short-wave radiation. Merry's data do not support a markedly more continuous distribution of aquatic species, although both the sampling strategy and the low diversity of aquatic species preclude a rigorous analysis. Fig. 14 shows the average number of groups of contiguous sites (stretches) at which terrestrial, aquatic and riparian species, present at between 6 and 15 sites, were recorded. If a species occurred at six spatially separate sites in the river system, then clearly it will be shown as occurring in six stretches; whereas if it occurred at six adjacent sites, it will be shown as occurring in one stretch. Analysis was confined arbitrarily to species occurring at 6 to 15 sites because, on the one hand, contiguity was inevitable for the very few species which occupied most sites and impossible for those many species which, at the other extreme, occupied only one site. With all three groups of species, aquatic, riparian and adventitious-terrestrial, the stretch:site ratio was between 55 and 65% for the types of plant and range of sites shown in Fig. 14.

Aquatic plants. The aquatic species showed some broad zonation (Fig. 15) except for *Elodea canadensis* and most of the *Potamogeton* spp. *E. canadensis* occurred at three widely scattered sites and *P. crispus, lucens, pectinatus, perfoliatus* and ×*salicifolius* at single, or two widely-scattered sites. At upland sites only *Potamogeton polygonifolius* and *Ranunculus omiophyllus* and *R. flammula* were found: Haslam (1978) records their presence in such upland streams draining peat. *R. flammula* extends further downstream in the Wye, to an altitude of 75 m, than might normally be expected. Although not found at the uppermost sites (>350 m) *Callitriche hamulata, Glyceria fluitans* and, to a lesser extent, *Myriophyllum verticillatum* occur only in the upper 100 km of the river; of these species only *G. fluitans* has a distribution extending for more than about 40 km. *G. fluitans* is more commonly associated with the margins of rather smaller streams and channels than the Wye and, although *M. alterniflorum* might have been expected, *M. verticillatum* is generally found in eutrophic and still waters – indeed there are very few records of its occurrence in Wales. *C. hamulata* is, however, commonly recorded from oligotrophic upstream reaches of rivers on resistant rocks, such as the Silurian and Ordovician formations of the upper Wye.

Veronica beccabunga, which may in different habitats be either submerged or emergent, is marginal and emergent in the Wye. It is widespread in the mid-reaches, principally on the marls of Old Red Sandstone. *Polygonum hydropiper,* another marginal plant, is also particularly common in the mid-reaches, although its distribution extends for about 200 km.

There are several species, both submerged and emergent, which, although extensive in their distribution, most commonly occur in the lowland reaches, and these include: *Ranunculus penicillatus, Myriophyllum spicatum, Mentha aquatica, Polygonum amphibium, Sparganium erectum, Lemna minor* and *Phalaris arundinacea.*

Ranunculus penicillatus, which occurred with *R. fluitans, R. peltatus* and hybrids, must be regarded as the dominant plant of the lower Wye where stands commonly extend across the river from bank to bank and where, during June and early July, its flowers cover the river surface, giving it the bizarre appearance of an unseasonal snow-fall. The quantitative aspects of its distribution, and its effects on water quality will be discussed later in the chapter (p. 42). *R. penicillatus* is, according to Haslam (1978), found in a range of aquatic habitats from 'swift streams on resistant rocks' to 'medium-sized mesotrophic to eutrophic streams of at least moderate flow, particularly on limestone-clay . . .'. However, it seems to be most abundant in wide rivers with moderate water velocities and depths and with an associated substrate containing sand.

Of the other species found in the lower reaches, *Lemna minor* deserves some mention. This floating species clearly needs an area of still water to grow. It only occurs in the Wye below the Lugg confluence, and its occurrence in the main river is dependent on production in marginal areas and back-waters of this tributary. The low flows during the survey period of 1976 no doubt increased its prevalence.

Haslam (1978) describes a survey of the R. Lugg and its tributaries in which more than 20 sites were sampled. Some species which she recorded (*Nuphar lutea, Schoenoplectus lacustris, Apium nodiflorum, Zannichellia palustris, Veronica anagallis aquatica, Sparganium emersum*) were not found by Merry in the Wye, perhaps the most surprising being *Sparganium emersum,* which was widespread. The low gradients of the Lugg catchment and its eutrophic character, account for some of the other occurrences e.g. *Nuphar lutea, Schoenoplectus lacustris.*

Riparian plants. The riparian species are listed in Appendix 1, those which are normally or frequently found adjacent to standing or flowing water, in contrast to the adventitious species being identified. Fig. 15 also shows the distributions of some of these riparian species, particularly the more abundant and widespread ones. Of those occurring at single or very few sites, and not shown in Fig. 15, the following, because of their rarity or particular distributions, deserve comment.

Carex pseudocyperus, Salix triandra and *Epilobium tetragonium* are predominantly found in South-Eastern Britain, and all are absent from the upper reaches of the Wye. In contrast, several species (*Pinguicula vulgaris, Mimulus moschatus, Hypericum elodes, Saxifraga stellaris, Epilobium neritoides*) are confined in Wales principally to the central, northern and western areas: these are almost invariably found only in the upland sites in the north-west of the catchment. *H. elodes,* most commonly found in bogs, is becoming rarer with the widespread drainage taking place in some areas. *Mentha rotundifolia,* recorded only near the mouth of the Wye, is restricted in Wales and South-West England to sites near the coast. *Montia*

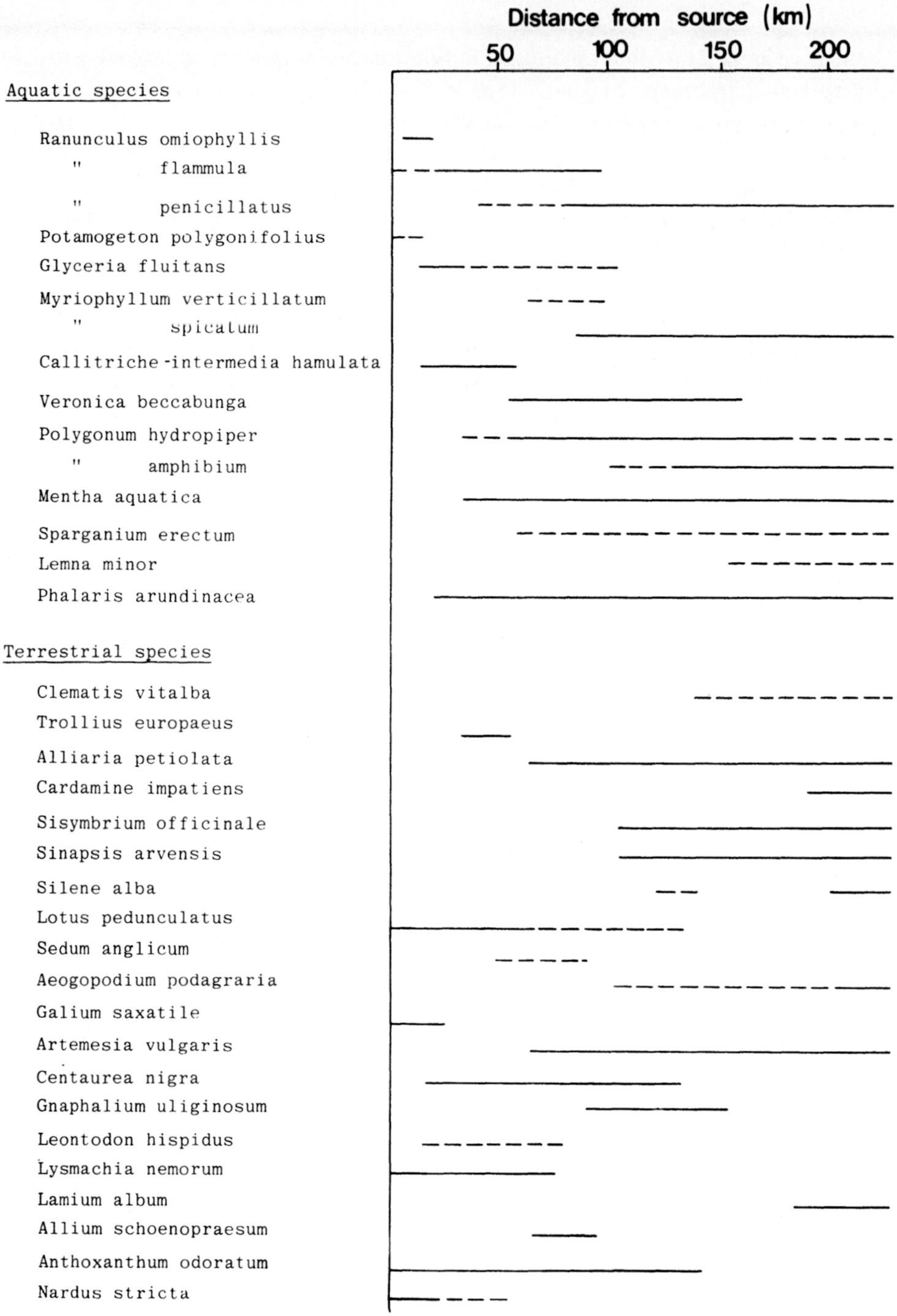

Distance from source (km)

50   100   150   200

Aquatic species

Ranunculus omiophyllis
    "   flammula
    "   penicillatus
Potamogeton polygonifolius
Glyceria fluitans
Myriophyllum verticillatum
    "   spicatum
Callitriche -intermedia hamulata
Veronica beccabunga
Polygonum hydropiper
    "   amphibium
Mentha aquatica
Sparganium erectum
Lemna minor
Phalaris arundinacea

Terrestrial species

Clematis vitalba
Trollius europaeus
Alliaria petiolata
Cardamine impatiens
Sisymbrium officinale
Sinapsis arvensis
Silene alba
Lotus pedunculatus
Sedum anglicum
Aeogopodium podagraria
Galium saxatile
Artemesia vulgaris
Centaurea nigra
Gnaphalium uliginosum
Leontodon hispidus
Lysmachia nemorum
Lamium album
Allium schoenopraesum
Anthoxanthum odoratum
Nardus stricta

Fig. 15 Distribution of many aquatic, riparian and terrestrial angiosperms in the Wye and on its banks.

36

Distance from source (km)

50   100   150   200

Riparian species

Caltha palustris
Ranunculus ficaria
Barbarea vulgaris
Brassica nigra
  "   rapa
Cardamine pratensis
Rorippa islandica
Viola palustris
Myosoton aquaticum
Sagina procumbens
Impatiens glandulifera
Potentilla erecta
Chrysosplenium oppositifolium
Lythrum salicaria
Epilobium hirsutum
Apium graveolens
Galium palustre
Succisa pratensis
Cirsium palustre
Eupatorium cannabinum
Senecio aquaticus
Wahlenbergia hederacea
Symphytum officinale
Mimulus guttatus
Veronica filiformis
Salix aurita
  "   viminalis
Narthecium ossifragum
Juncus bufonius
  "   effusus
  "   squarrosus
Carex remota
  "   panicea, curta &
               echinata
Eleocharis palustris
Alopecurus geniculatus
Molinia caerulea

37

sibiria, an introduction from America, is not common in Wales, although it is widespread in Northern England and Scotland; on the Wye it is found frequently in the middle reaches.

Of the riparian species shown in Fig. 15, *Wahlenbergia hederacea, Narthecium ossifragum, Juncus squarrosus* and *Carex panicea, C. curta* and *C. echinata* are confined to upland sites, not being recorded in the catchment below 350 m. The distributions of *W. hederacea* and *N. ossifragum* in Wales are principally restricted to the uplands of western and central areas and *J. squarrosus* is not common in the south-east of the Principality: the other species are more widely distributed but are normally associated with wet, upland, acid sites. Other upland species, such as *Molinia caerulea, Viola palustris, Cardamine pratensis, Cirsium palustre, Salix aurita* and *Juncus effusus* extend further down the catchment.

Another group of species, characteristic of the uplands, appears within 50 km of the source and above an altitude of 150 m:

Caltha palustris	*Galium palustre*
Ranunculus ficaria	*Succisa pratensis*
Sagina procumbens	*Carex remota*
Chrysosplenium oppositifolium	*Alopecurus geniculatus*

All are fairly widely distributed in Britain, and their location in the Wye catchment will be determined principally by local site factors. *Senecio aquaticus,* too, which in Britain has been recorded at an altitude of about 400 m, extends up the Wye catchment from the downstream sites to a height of about 240 m.

Comparatively few species are confined to the mid-reaches of the Wye (*Apium graveolens, Mimulus guttatus, Juncus bufonius, Eleocharis palustris*). A further species, *Epilobium hirsutum,* although extending downstream, is most common in this zone. It is absent from the upstream stations in the catchment and generally uncommon in Central and West Wales – in contrast with its widespread distribution in the uplands of Northern England. *Apium graveolens* is generally restricted to the coasts, and there is only one previous inland record of this species in Wales. *M. guttatus,* the Monkey flower, a native of North America, first appeared in Britain as a garden escape in Wiltshire in 1830. A plant of river banks and wet meadows, it spread particularly rapidly and was widespread a century later. Its restriction to the middle reaches of the Wye is rather surprising, for it occurs extensively in the lowlands of Southern Britain.

Several of the riparian species shown in Fig. 15 are confined principally to the lower part of the catchment:

Barbarea vulgaris	*Impatiens glandulifera*
Brassica nigra and *B. rapa*	*Lythrum salicaria*
Rorippa islandica	*Eupatorium cannabinum*
Myosoton aquatica	*Symphytum officinale*
Veronica filiformis	

B. nigra (Black mustard), a lowland species not recorded above about 300 m in Britain, has a scattered distribution in Wales where it occurs principally on stream banks and sea-cliffs: it has probably been cultivated from the Iron Age for its oily seed. *R. islandica* and *M. aquaticum* are common in Central and West Wales so their absence from the upper reaches of the Wye may be associated with broad distributional factors rather than site characteristics. *I. glandulifera*, a species introduced to Britain from the Himalayas in 1839, has extensively colonised river banks comparatively recently and it still has only a scattered distribution in much of Wales. *L. salicaria* occurs in Wales most abundantly around the West and South-East coasts and it is absent from some inland areas: its restriction in the Wye catchment to the lowest 100 km (apart from one site in the mid-reaches) fits this general distributional pattern.

Terrestrial plants. Most species recorded in the margins of the Wye are found more commonly in habitats other than river-banks, and their occurrence in this marginal habitat is evidence not only of its suitability for colonisation and growth, and perhaps propagation, but also the likely proximity of a more usual habitat supplying an abundance of propagules. Thus the occurrence of the grasses *Nardus stricta* and *Anthoxanthum odoratum* and the herbs *Trollius europaeus, Lotus pedunculatus, Galium saxatile, Genista tinctoria* and *Centaurea nigra* in the upper reaches might be expected from their common occurrence on heaths, moors and damp pastures of the uplands and that of *Sinapsis arvensis* and *Sisymbrium officinale*, common weeds of arable land, in the lower reaches (Fig. 15). Similarly *Anemone nemorosa*, normally a woodland species, is prevalent in those reaches where banks are extensively shaded by trees.

The distributions of some species are more directly explicable from the geology of the catchment. Thus the calcicoles *Clematis alba, Cardamine impatiens, Diplotaxis muralis* and *Silene alba*, as well as *Acer campestre*, which avoids acid soils, are largely confined to the downstream limestone reaches. In contrast, *Gnaphalium vuliginosum*, which prefers sandy soils, occurs on the marls of the Old Red Sandstone.

The distributions of many species in the Wye cannot be linked with local site factors, but some are consistent with their broader geographical distributions in Britain. It must be remembered that the Wye is a linear habitat about 250 km in length and interpretation requires reference to biogeographical factors. Many species with a predominantly south-eastern distribution in Britain were found on the Wye, almost invariably in the lower and middle reaches:

Coronopus didymus	*Thalycrania sanguinea*
Armoracia rusticana	*Galium cruciata* and *G. mollugo*
Reseda lutea	*Arctium lappa.*

Other species (*Ononis spinosa, Lathyrus sylvestris, Trifolium ornithopodiodes* and *Viccia lathyroides*), of generally similar distribution (E., S.E. or N.E.) in

Britain, but found in South-West England and Wales principally around the coasts, were also recorded, but curiously in the middle rather than lower reaches of the Wye. *O. spinosa, T. ornithopodiodes* and *V. lathyroides* are rare in Wales.

A few species, absent from the upper stations of the Wye (*Alliaria petiolata, Artemesia vulgaris, Campanula latifolia* and *Lamium album*), are generally widespread in England and Wales but are not found in the more western regions.

In contrast, there are a few species with western distributions in Britain which were found in the Wye only at sites in the upper and middle reaches, the most interesting being *Viccia orobus, Allium schoenopraesum* and *Sedium rosea* and *S. anglicum. S. rosea* has been rarely recorded in Wales except in the north-west. *A. schoenopraesum*, the chive, has become more widespread over the past century in the middle reaches of the Wye, particularly on the emergent rocks and boulders around Builth Wells, but it has been found at relatively few sites elsewhere in Britain. A further species, *Potentilla rupestre*, recorded by Merry at one site, has only been found in Britain within central Wales.

Species with other patterns of distribution in Britain have also been found on the Wye, their local distributions within the catchment generally, but not invariably, being consistent with their broader distributional patterns. Species rare in Wales with such distributions are:

Chamaemelum nobile	(Southern)
Polygonatum multiflorum	(Southern)
Myosotis sylvatica	(North Eastern)
Festuca vivipara	(North Western)

Vertical distribution of plants on river banks (Fig. 16)

In the upland sites (Fig. 11) bryophytes comprised an important component of the flora and several species, e.g. *Marsupella emarginata, Scapania undulata, Hypocomium flagallare*, occurred above and below water level. Species typical of upland grassland, such as *Deschampsia caespitosa, Rhytidiadelphus squarrosus, Festuca ovina* and *Galium saxatile*, were confined to the drier hummocks where soils were relatively well drained.

In contrast, lowland sites were dominated by angiosperms many of those at the water's edge, such as *Juncus bufonis, Rorippa islandica, Polygonum amphibium, Sparganium erectum* being truly riparian. Large areas at many sites were covered by the tall herbs (e.g. *Chrysanthemum vulgare, Myosoton aquaticum* and *Artemesia vulgare*), several being adventitious to the riparian habitat. Alternatively the grasses such as *Agrostis stolonifera, Dactylis glomerata, Holcus lanatus, Agropyron repens* and *Arrhenatherum elatius* dominated the banks: these species were present at many upland sites too.

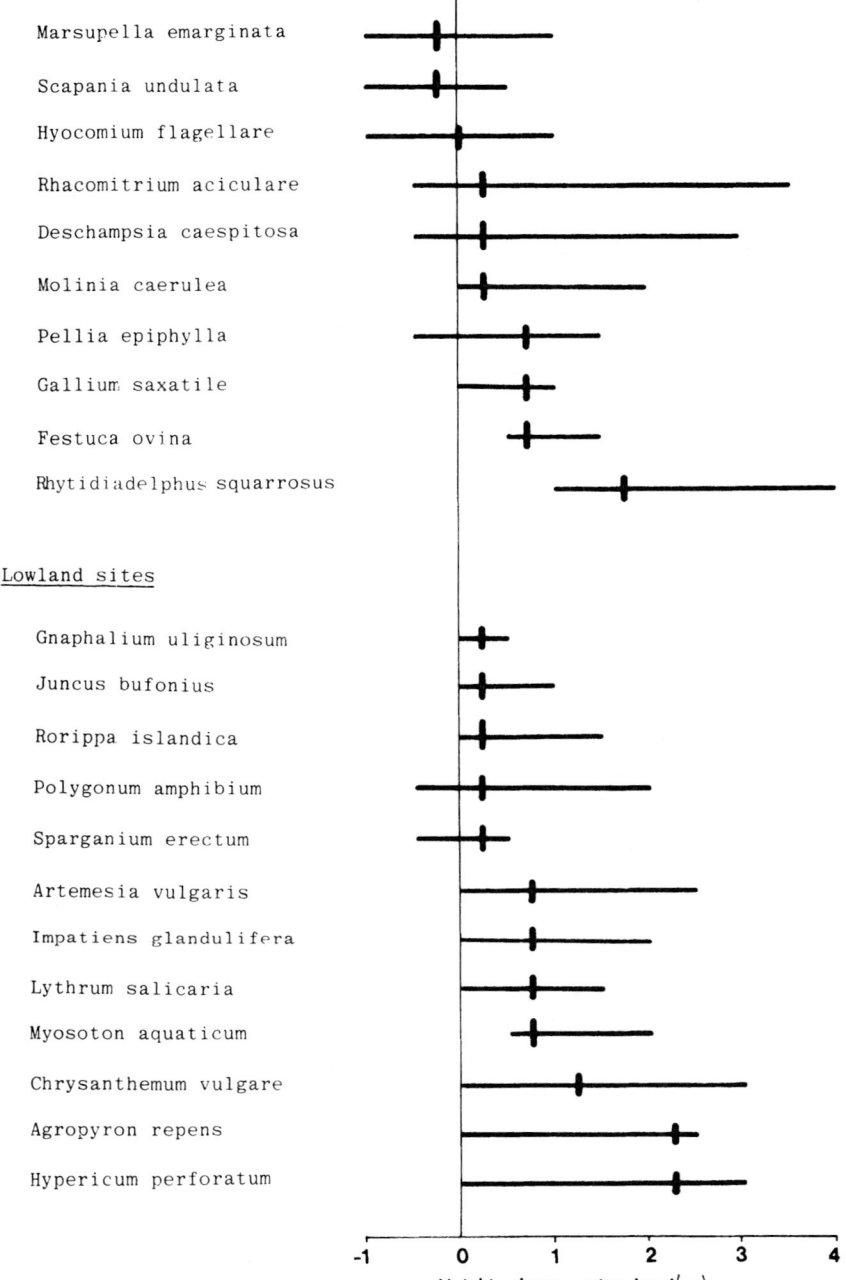

Upland sites

Marsupella emarginata

Scapania undulata

Hyocomium flagellare

Rhacomitrium aciculare

Deschampsia caespitosa

Molinia caerulea

Pellia epiphylla

Gallium saxatile

Festuca ovina

Rhytidiadelphus squarrosus

Lowland sites

Gnaphalium uliginosum

Juncus bufonius

Rorippa islandica

Polygonum amphibium

Sparganium erectum

Artemesia vulgaris

Impatiens glandulifera

Lythrum salicaria

Myosoton aquaticum

Chrysanthemum vulgare

Agropyron repens

Hypericum perforatum

Height above water level (m)

Fig. 16 Vertical distribution of bryophytes and angiosperms on the banks of the Wye at a typical upland and lowland site. Vertical bar represents height of maximal abundance.

41

Growth of Ranunculus penicillatus and its effects on river conditions

Downstream of Hereford *Ranunculus penicillatus,* the water crowfoot, grows abundantly during the spring and early summer, the most profuse growths generally occurring between Ross-on-Wye and Monmouth. Hutton (1930) described similarly profuse growths in the early years of this century – before the widespread use of artificial fertilisers to which such growths of aquatic plants are sometimes attributed – and perceptively he questioned their adverse effects on salmon.

Brooker et al. (1978) have studied the seasonal growth of water crowfoot at a few sites (W53, W54 and W55, for site locations see Fig. 2) during the period 1975–77. Perceptible growth normally starts late in April, and flowering and die-back generally occur at the end of June or beginning of July, although it may be later in some years. There is considerable inter-year variation of maximum biomass: at Kerne Bridge (W54) in 1976 about 2 kg fresh wt/m^2 was recorded, which compares with maximum yields of 4–7 kg/m^2 found in the most productive submerged plant communities (Westlake 1963). Brooker et al. complemented their intensive studies in limited reaches by more extensive cover estimates, and these also demonstrated the wide variation in growth from year to year, peak cover between Ross and Monmouth being 17, 36 and 9% in 1975, 1976 and 1977 respectively. Converting these cover estimates into weight, these authors concluded that the river supported about eight times as much plant material in 1976 as in 1977.

It seems that the maximum biomass (PB: kg fresh wt/m^2) and the percentage cover (C) at sites in the reach from Ross to Monmouth may be predicted in any year from average daily river flows (F: m^3/s) during the growth period (April–June) using the following regression equations:

$$PB = 2.64 - 0.05F$$
$$C = 54.0 - 1.03F$$

Brooker et al. (1978) have also subjectively ranked peak annual plant-growth for the period 1968–75 from information given in the Wye River Authority Fisheries Reports and shown its negative relation with average flow in April–June.

Such substantial plant-growth clearly has many consequences to the ecology, water-quality and management of the river. Two effects have been studied in some detail: (a) dissolved oxygen concentration of river water and (b) river hydraulics and frictional losses.

Dissolved oxygen. For long periods during 1975, 1976, 1977 and 1978 submersible dissolved oxygen and temperature recorders were located at several sites in the lower catchment, the most complete record being from Kerne Bridge (W54).

During the winter months dissolved oxygen concentrations varied little (11–13 mg/l, being close to the air saturation value at the prevailing temperatures (2–4 °C). During the late spring and early summer, when *Ranunculus* was growing

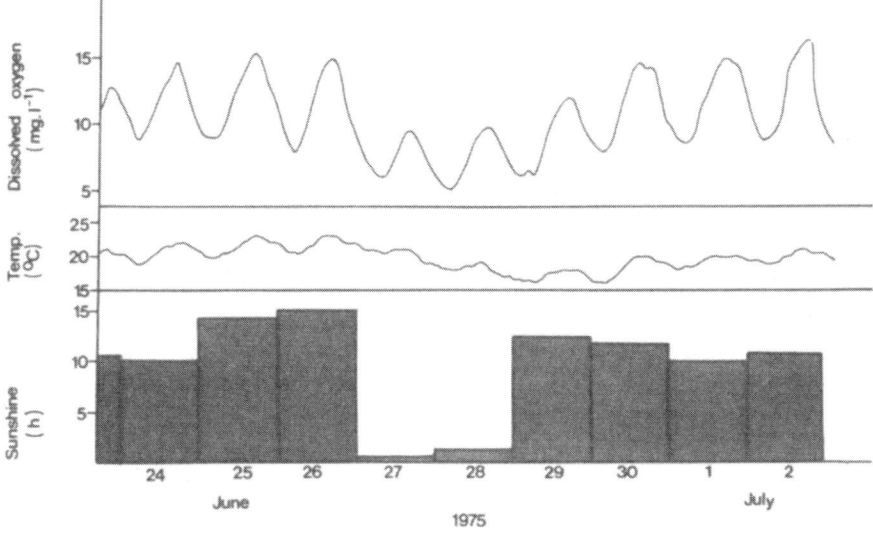

Fig. 17 Dissolved oxygen concentrations, temperature and hours of sunshine in the lower Wye (W54) during late June and early July 1975.

and when flows were reduced, there were substantial daily fluctuations, generally ranging from 5 to 15 mg/l (Fig. 17). Such diurnal changes were considerably modified by meteorological factors, particularly solar radiation.

It was possible to calculate the gross primary oxygen production from daily changes in oxygen mass within the river, assuming that oxygen concentrations behaved similarly throughout reaches where oxygen recorders were located. Such estimates ranged from about 1 to 13 g $O_2/m^2/d$, community respiration being within the same range – these values are similar to estimates from several other rivers, lakes and ponds.

During the summer of 1976 low flows and high temperatures occurred when the very large growths of water crowfoot, promoted by a dry spring, were dying and decaying (Brooker et al. 1977). This combination of factors caused dissolved oxygen concentrations to fall during darkness to about 0.5 mg/l (6% saturation). Substantial mortalities of salmon occurred during this period, over 500 corpses being recovered from the river, as a consequence of both low dissolved oxygen concentrations and maximum temperatures around 27°C.

There are previous examples of mortalities of salmon during periods of low flow and high temperature: for example, about 350 salmon corpses were found during May–September 1934 when afternoon water temperatures at Symond's Yat occasionally exceeded 25°C. In the previous year about 200 corpses were similarly observed during summer heat-waves. In contrast, June 1940 was reported as the hottest period for about 50 years, and yet few salmon seem to have died.

43

River hydraulics. By increasing frictional resistance to water flow, plants can increase water depth and lead to flooding. Less dramatically they can interfere with flow-gauging procedures by altering flow-depth relationships at gauging stations throughout their growth cycle. Brooker et al. (1978) made contemporaneous measurements of plant biomass and river slope at Belmont (W48) on the Wye and at Lugwardine (L1) on the R. Lugg. From slope and other site characteristics, Manning's coefficient of friction (n) was calculated and shown to be correlated with plant biomass - after the effects of variations in flow, to which n is related, were eliminated. Estimates of n varied, during the growth cycle, between 0.02 and 0.25 at Lugwardine and 0.02 and 0.10 at Belmont: these compare with values in most natural streams of between 0.025 and 0.075.

This increased frictional resistance caused by plant growth is not likely to pose a problem in the flooding of land, the river channel of the R. Wye being sufficiently large to accommodate the small increases in water depth, calculated to be less than a few centimetres at Redbrook (W51) at normal summer flows and with peak plant biomass. In the Lugg, a more sluggish and channelised river, increases in depth resulting from plant growths may be about 15 cm, still substantially less than the 30 cm difference in depth recorded between open and weedy reaches in chalk streams in the south of England (Dawson 1978).

With respect to flow gauging, it has been shown that the flow can be estimated at gauging stations on the Wye more accurately from stage height if plant biomass is taken into account, particularly in years of heavy growth. However, the improvement in accuracy must be offset against the cost of estimating the biomass and alternative procedures are available to improve accuracy, the most general being frequent re-calibration of the stage height–flow relationship.

Algae

The study of algae in large river systems has never been as fashionable as that in lakes and reservoirs, although occasional studies have been reported (Lack 1971) in the U.K. The reasons for this comparative neglect range from the simple practical problems of travelling long distances for sampling to the difficulties of interpreting changing populations in both spatial and temporal terms where many species may grow both on a static benthic substrate and suspended within moving water.

Recent algal studies have been undertaking in the R. Wye by Dr J. E. Furet (1979) and these represent the first detailed descriptions, apart from routine observations carried out by the Welsh Water Authority in the lower Wye (W59) in relation to the monitoring of water quality at Monmouth and a summer survey of 'planktonic' algae of the lower Wye (Mr F. H. Jones, personal communication. This section is based principally on Dr Furet's studies during 1975–76 in which water samples were collected, generally at two-week intervals, at ten sites on the main river and at one site on each major tributary (Elan, Ithon, Irfon and Lugg): these sites coincided with those at which water quality was determined (see Chapter 2).

Within the samples two broad groups of algae could be recognised, those which, referred to as 'bloom species', grow planktonically and others which are there adventitiously, having been derived largely from benthic growth. There were however several species which probably grew well in both their benthic and planktonic phases. The algae which occurred abundantly in summer blooms were chlorophyceans (*Scenedesmus* and *Ankistrodesmus*) and several small centric diatoms (*Cyclotella, Stephanodiscus, Thalassiosira*). In contrast with other rivers studied in the U.K. where *Stephanodiscus hantzschii* is dominant, *Scenedesmus* is commonly the most abundant species in the R. Wye.

Dr Furet concluded that the Wye catchment divides naturally, on algal distributions, into three zones.

1) The *upper zone*, low in calcium (<10 mg/l), which extends from the headwaters to Builth Wells and includes the R. Elan (but not the R. Ithon), has less than 8 μg/l of chlorophyll *a* derived from algae suspended in the water, and no seasonal peak in chlorophyll concentrations. Despite the overall lack of seasonality, different algae are prevalent at different times of year. *Meridion circulare* occurs during the winter, whereas *Scenedesmus* spp., *Ankistrodesmus falcatus, Rhoicosphaenia curvata* and *Navicula cryptocephala* are dominant only during the summer. Several benthic diatoms, such as *Synedra ulna, Diatoma vulgare, Nitzschia palea* and *N. amphibia,* are abundant and seem to cause appreciable reductions in the concentrations of nitrate and dissolved silica.

2) The *middle zone*, extending from Builth Wells to Erwood (W22–28) and including the R. Irfon, with calcium concentrations usually within the range 10–25 mg/l, generally has chlorophyll *a* concentrations below 20 μg/l with peaks in June–July and November. As in the upper reach, *Meridion circulare* occurs during the winter and bloom species (*Scenedesmus* spp., *Cyclotella meneghiniana, Thalassiosira fluviatilis* and *Carteria* spp.) during the summer. Several benthic diatoms are abundant during the spring (*Melosira varians, Surrirella ovata, Navicula viridula, Ceratoneis arcus, Diatoma vulgare* and *Pinnularia viridula*), and their growth is coincident with a decrease in dissolved silica.

3) The *lower zone*, downstream of Hay-on-Wye (but including the R. Ithon which is extraordinarily rich in algae), has peak chlorophyll *a* concentrations well above 20 μg/l. The R. Lugg occasionally reaches chlorophyll *a* concentrations around 100 μg/l. As in the middle and upper zones there is a group of 'bloom species' during the summer; of these *Monorhaphidium gracilis* and *M. tortile* and *Oocystis crassa* increase their proportional abundance in this lower zone.

Of the tributaries, the R. Elan is surprisingly poor in algae in view of its reservoir source. Nevertheless it contains higher concentrations of *Tabellaria flocculosa* and *T. fenestrata* than are found elsewhere in the catchment. In contrast the R. Ithon, draining a catchment of low relief, is very rich in algae and contrasts markedly with the R. Wye at their confluence. Several of the species abundant in the R. Ithon, such as *Diatoma vulgare, Cocconeis placentula* and *Nitszchia* spp., also seem suited to conditions in the lower Wye where they become even more abundant. The R. Irfon although generally similar to the upper Wye is rich in the Chlorococcales, particularly species of *Scenedesmus*. The Lugg has the highest algal concentrations in the Wye catchment, and, like the Ithon, it drains an area of low relief: it is, however, much richer in plant nutrients, draining fertile farmland, much of it arable, in its lower course.

Studies during the summer of 1980 (Mr F. H. Jones, personal communication) suggested that, in the lower R. Wye, increases in densities of 'bloom species', particularly *Scenedesmus* which represented about 90% of total cell numbers (about 10^5/ml) in June, were synchronous. Nevertheless the total number of cells flowing down the river increased between four-fold and eight-fold over a reach of about 100 km (W45 to 62) with a residence time of seven days: this increase represents a cell division every two days. In contemporaneous studies in the R. Lugg and contrasting with observations of Furet (1979), all concentrations were much lower than in the R. Wye, peak concentrations being 1.3×10^3 and 10^5 cells/ml respectively.

Plant distributions in the Elan Valley reservoirs

Marginal macrophytes. Although extensive surveys of the reservoir macrophytes have not been undertaken, Merry and Slater (1978) described the marginal bryophytes and angiosperms colonising the draw-down zone of Caban Coch Reservoir in the Elan Valley (Fig. 2) during the summer of 1975, following one of the driest nine-month periods of this century. These authors concluded that the reservoir margins, being an ephemeral and irregularly occurring habitat, were inhabited principally by those riparian species which could colonise the exposed surfaces rapidly. They distinguished associations containing species characteristic of (1) acidic upland pastures (e.g. *Agrostis stolonifera, Festuca ovina, Hypnum cupressiforme, Nardus stricta, Polytrichum commune, Sphagnum recurvum*); (2) strandline (e.g. *Polygonum hydropiper, Gnaphalium uliginosum, Betula pubescens, Juncus effusus*); and (3) open mud (e.g. *Juncus bufonius, Leptodictyum riparium, Peplis portula, Polygonum persicaria, Pseudephemerum nitidum, Solenostoma obovata*).

In addition, there were some aquatic and semi-aquatic species, including *Callitriche* spp., *Drepanocladius fluitans, Glyceria fluitans* and *Pellia epiphylla*.

Algae. The only published studies on algae are those of Round (1956, 1957), who compared the phytoplankton of the newly filled Claerwen Reservoir with that of the long established Craig Goch and Caban Coch reservoirs and described the benthic algae at three sites, in Craig Goch, Caban Coch and Dol-y-mynach (a small reservoir between Claerwen and Caban Coch). Clearly, with land-use changes affecting nutrient loadings and increases in reservoir age and utilisation, the algae of this system may now be different, and a detailed investigation of their distribution is overdue.

About 35 species of phytoplankton were recorded from the reservoirs, the majority of species belonging to the Chlorophyta (Table 10) and the commonest genus being *Staurastrum*. Some species, although found in the plankton, were probably incidental captures resulting from the disturbance of the sediments in these exposed reservoirs, e.g. *Ulothrix* sp., *Hormidium* sp., *Micrasterias* spp., *Pleurotaenium trabecula*.

Table 10 Principal species of algae recorded in the plankton of the Elan Valley reservoirs, 1952–55.

Chlorophyta	Chrysophyta
Ulothrix sp.	*Dinobryon sertularia*
Hormidium sp.	*Mallomonas longiseta*
Staurastrum paradoxum	
Staurastrum anatinum	**Cyanophyta**
Staurastrum jaculiferum	*Anabaena*
Xanthidium antilopaeum	
Xanthidium armatum	
Closterium kutzingii	**Bacillariophyta**
Pleurotaenium trabecula	*Tabellaria flocculosa*
Micrasterias rotata	*Tabellaria fenestrata*
Micrasterias denticula	*Fragilaria* spp.
Micrasterias truncata	
Hyalotheca dissiliens	**Pyrrophyta**
Gymnozygon moniliformis	*Peridinium willei*
Desmidium swartzii	
Cosmocladium saxonicum	
Spirogyra sp.	
Zygnema sp.	
Mougeotia sp.	

In general, more species were recorded from Caban Coch and Craig Goch than the, then newly filled, Claerwen Reservoir, but the flora, overall, was typical of nutrient-poor water bodies. The absence of certain species, such as *Asterionella formosa, Fragilaria crotonensis* and *Rhizosolenia euensis,* reflects the extreme oligotrophy of these reservoirs. The high proportion of desmid species in characteristic of natural lakes in Wales, the Lake District and Scotland, and the reservoirs in the Elan Valley seem to support a very similar flora to these natural water bodies.

Seasonal variation in the numbers of the common species showed two patterns, some species having a single annual growth maximum and the others two annual maxima (Table 11).

47

Table 11 Typical periods of maximum numbers of common phytoplankton in the Elan Valley reservoirs.

	Craig Goch	Caban Coch	Claerwen
Spring	*Dinobryon*	*Dinobryon* *S. paradoxum*	*Staurastrum paradoxum*
Summer		*S. paradoxum*	*S. paradoxum*
Autumn	*Gymnozygon moniliformis*	*S. paradoxum* *G. moniliformis*	*S. paradoxum*
Winter	*S. paradoxum*	*S. paradoxum*	*S. paradoxum*

Table 12 The most frequently occurring benthic algae in the Elan Valley reservoirs, 1953-55.

Nitzchia dissipata	*Pleurotaenium trabecula*
Neidium iridis	*Euastrum ansatum*
Pinnularia gibba	*Closterium didymotocum*
Nitzchia palea	*Closterium pritchardianum*
Anomoeoneis exilis	*Closterium lunula*
Navicula pupula var. *capitata*	*Closterium intermedium*
Neidium affine var. *amphigomphus*	*Penium libellula* var. *intermedia*
	Penium naviculata
	Penium spirolostriatum
	Tetmemorus brebisonii
	Oscillatoria irrigua
	Oscillatoria splendida
	Merismopedia elegans
	Euglena mutabilis
	Euglena spirogyra

In contrast with the phytoplankton, the benthic algae chiefly comprised diatoms. The principal species of diatoms and other algal groups are shown in Table 12; of the latter only *Oscillatoria irrigua* was abundant. The benthic diatoms showed distinct seasonal patterns of abundance, with increases in cell numbers generally being recorded during the period April June.

Vegetation of Llangorse Lake

In addition to the Elan Valley reservoirs, the Wye catchment contains Llangorse Lake (Fig. 18), the largest lake in South Wales with an area of 150 ha. Principally because of the rarity of some of its riparian plant species, such as *Nymphoides peltata, Rumex maritimus* and *Ruppia maritima,* it has been scheduled as a Site of Special Scientific Interest. There are historical records of the lake having algal

48

blooms since the twelfth century (Jones & Benson-Evans 1974) and substantial growths of submerged macrophytes since the late ninteenth century, phenomena associated with the reservoir's high nutrient status and shallow character, the average depth being only about 2.6 m. Dramatic changes have occurred in recent years which seem related not only to the hyper-eutrophy of the water, associated with the increasing discharge of sewage effluent and application of fertilisers within the catchment, but, perhaps more importantly, to the recent escalation in water-based recreation, and most particularly, power-boating.

Algae. The algae of the lake have been described by Jones & Benson-Evans (1974). The normal seasonal sequence of diatoms in the late winter and early spring, succeeded by green algae during the early summer, blue-green algae during the late summer and autumn and diatoms again in the late autumn, is generally followed in Llangorse. *Asterionella formosa, Stephanodiscus dubius, S. hantzschii* and *S. astrea* are generally the most abundant diatoms, *Ankistrodesmus falcatus, Dictosphaerium ehrenbergianum* and *Scenedesmus quadricauda* frequently the most abundant green algae, and *Anabaena flos-aquae, Aphanizomenon flos-aquae* and *Microcystis aeruginosa* the dominant blue-green algae: *Tribonema bomyanium* (Xanthophyceae) is sometimes common during the summer months. On occasions, the concentration of *Microcystis* exceeds 10^8 cells/l and, at such times, chlorophyll *a* concentrations are greater than 400 μg/l.

During the spring diatom bloom, concentrations of silica occasionally decrease from 12 to 1 mg/l, at which concentration growth is frequently limited.

Periodic mortalities of fish, such as that of roach in 1970, occur, and it has been suggested that toxins produced by certain blue-green algae in the lake, particularly *Microcystis,* are implicated.

Macrophytes. The submerged macrophytes have changed dramatically in recent years (Cragg et al. 1980), and Table 13 shows the decline in the number of species from 11 in 1964 to only one, *Zanichellia palustris,* in 1977, this remaining species being confined to areas generally less than 2 m in depth (Fig. 18). The maximum density of *Z. palustris* in 1977, except for a few small patches, was only about 20 g fresh wt/m² which compares with over 2 kg fresh wt/m² of *Ranunculus* in the lower Wye (p. 42). It has been suggested that the paucity of macrophytes is caused by the reduced light intensity associated both with high densities of algae and resuspended sediment, and possibly by the physical damage to young shoots caused by power-boating.

The margins of the lake have characteristic associations of marsh, sedge swamp, reed swamp and floating-leaved plants (Table 14), and their distribution is shown in Fig. 18. In 1977 when the lake was last surveyed (Cragg et al. 1980), the largest stands of emergent and floating-leaved vegetation were located primarily in the sheltered western and southern shores, this vegetation being typified by *Phragmites australis, Typha latifolia* and *Nuphar lutea,* all of which might be expected in

Table 13 The presence (+) or absence (–) of submerged macrophytes in Llangorse Lake between 1964 and 1977 (Those marked with an *, although not present in the lake, are found in the Afon Llynfi which flows into and out of it).

Species	1964	1972	1973	1977
Chara sp.	..	+	–	–
Elodea canadensis	+	+	+	–*
Zannichellia palustris	+.	+	+	+
Potamogeton berchtoldii	+	–	–	–
Potamogeton crispus	+	+	+	–*
Potamogeton lucens	+	+	–	–*
Potamogeton natans	+	–	–	–*
Potamogeton pectinatus	+	+	+	–
Potamogeton perfoliatus	-	+	+	–
Potamogeton pusillus	-	+	+	–
Lemna trisulca	+	–	–	
Myriophyllum spicatum	+	+	+	–*
Ranunculus aquatilis	+	–	–	–
Ranunculus circinatus	–	+	+	–
Ranunculus trichophyllus	+	–	–	–

highly eutrophic lakes (Cragg et al. 1980). On the more exposed northern and eastern shores, marginal zones were narrower and *Nymphoides peltata* and *Polygonum amphibium* were dominant. Behind the areas of reed-swamp, associated with free-standing water, the marsh and sedge vegetation consisted of four distinctive types. Peripheral to the reed-swamp there were generally zones of *Iris pseudacorus* and *Carex* spp., and these were fringed by copses of *Alnus glutinosa* and wet meadows dominated by *Juncus effusus, J. inflexus* and *J. acutiflorus.*

Table 14. Zone of marginal vegetation at Llangorse Lake (see Fig. 18).

Association	Society
Pasture and meadow	Grasses
Marsh	*Juncus*
	Alnus
Sedge swamp	*Carex*
	Iris
Reed swamp	*Phragmites*
	Schoenoplectus
	Typha
	Equisetum
Floating-leaved plants	*Nuphar*
	Nymphaea
	Nymphoides
	Polygonum

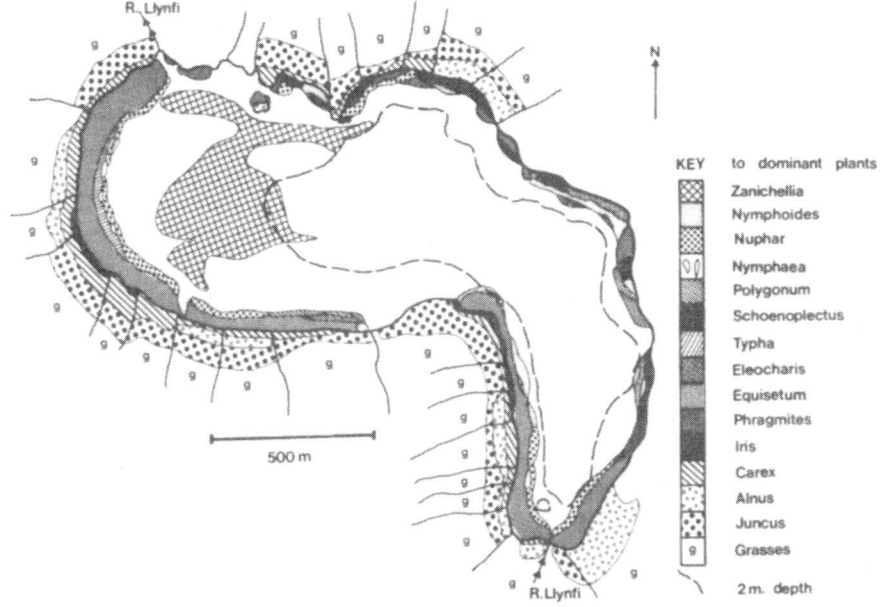

Fig. 18 Submerged and marginal macrophytes of Llangorse Lake in 1977.

The effects of variation in agricultural practice, particularly fertiliser application and grazing regime, on marsh and sedge vegetation are pronounced, as is shown by discontinuities at field boundaries (Fig. 18). Cattle extend their influence to emergent and floating-leaved plants and frequently eat *Nuphar lutea, Nymphaea alba, Typha latifolia* and *Schoenoplectus tabernaemontani*. Although not as pronounced as for submerged species, some recent changes in marginal vegetation have been noted, particularly a retreat of *Phragmites* on the eastern edge of the lake, attributed to mechanical damage from wave action associated with power-boats, and a concomitant expansion of the *Nuphar lutea* zone. There has also been a reduction in the abundance of *Schoenoplectus lacustris* and *S. tabernaemontani* and of *Equisetum palustris*.

Plate 2 Dense growth of *Ranunculus penicillatus* in the River Wye at Ross-on-Wye (W54).

4. Macroinvertebrates

Macroinvertebrate distributions within the river

The study of the natural history of the aquatic invertebrates – unlike investigations of other organisms, such as plants, which are easily observed by amateur naturalists – is constrained by their habit and habitat and, in consequence, there are few early reports of the macroinvertebrate fauna of the R. Wye. One of the first referred to the largest aquatic invertebrate in the Wye catchment, the crayfish, which was probably introduced to the R. Irfon from the R. Usk in the late eighteenth century (Jones 1805). Other information on invertebrates is generally restricted to incidental reports of adult insects, chiefly mayflies and stoneflies, the Wye being a haunt of collectors over a considerable period.

Hellawell (1971c) principally reporting fish studies undertaken in the mid-1960s in the Llynfi and Lugg, commented upon the distribution of crayfish in these rivers and upon the invertebrate component of the food of chub in the Wye catchment, but the first extensive survey of macroinvertebrates was undertaken in 1970 by Nature Conservancy Council (Ratcliffe 1977). This formed part of a broader ecological assessment of the river and provided a general description of the invertebrate fauna, noting characteristic upland and lowland communities. Several rare aquatic species were listed, notably mayflies – *Emphemera lineata* Ent., *Brachycercus harrisella* Curt. and *Potamanthus luteus* (L).

It was not until 1975 that a comprehensive study of invertebrates in the R. Wye and some of its major tributaries was carried out, and the following account is based largely on information collected during 1975–77 by several workers. General surveys (Brooker & Morris 1980a) were undertaken in order to describe the distribution and relative abundance of macroinvertebrate communities, principally in the main river, these data being supplemented by specific studies on the R. Elan (Paull 1978), on the distribution of crayfish in the upper Wye catchment (Lilley et al. 1979) and of adult aquatic insects in the river corridor, particularly at Newbridge-on-Wye (Morris 1981).

The general surveys included 12 sites on the main R. Wye, between 7 and 220 km from the source, and one site each on the impounded R. Elan and the R. Irfon (Fig. 2). All samples were collected from riffles – relatively shallow, fast-flowing reaches – although the size and mean depth of the river varied considerably from site to site. Pools were not routinely investigated, but Wisniewski (1978), who studied three sites in the upper Wye, concluded that the fauna of riffles and pools was qualitatively and quantitatively similar. Work in other rivers and streams, such as those in the uplands of Northern England, suggests that this is not always the case. Such heterogeneity may restrict the assessment of animals characteristic of habitats other than riffles, particularly in the lower reaches of the Wye which are predominantly deep and slow-flowing.

Generally each site in the Wye catchment was visited on six occasions – March and September 1975 and July and September 1976 and 1977. Replicate samples, four at each site, were collected with a quantitative sampler (Neill 1938), which is likely to have retained larger specimens of all invertebrates except Protozoa, Rotifera and Nematoda.

Many of the aquatic invertebrates cannot be identified easily, if at all, to species, especially in their juvenile stages, and the level of identification which was achieved varied considerably between groups. Some taxa, which could only be formally identified to family or genus, were further separated on morphological features to provide discrimination between organisms.

227 taxa were recorded from the six surveys of the Wye (Appendix 3) and were distributed between the major groups as follows:

- 44% Diptera (mainly Chironomidae),
- 9% Ephemeroptera,
- 9% Trichoptera,
- 8% Oligochaeta,
- 7% Plecoptera,
- 7% Coleoptera and
- 5% Mollusca.

Some of the invertebrates collected from these surveys of the Wye and regarded as rare in the U.K. included *Potamanthus luteus* (mayfly), *Normandia nitens* (beetle) and *Brachyptera putata* (Newman) (stonefly, collected as an adult). Many of the midge larvae, of course, are not well recorded in the U.K. because of the taxonomic uncertainties of the group, but *Rheotanytarsus distinctissimus* Br., collected as an adult at Newbridge-on-Wye, has not been recorded previously from the U.K. (Morris 1981).

The macroinvertebrate species list of the Wye catchment forms only a small proportion of the total number (>3500) of species recorded from freshwater habitats in the U.K. but is generally similar to that collected from other relatively base-poor river systems, like the R. Endrick in Scotland (Maitland 1966). Overall, most taxa (133) in the Wye were recorded from the most downstream site sampled and fewest (64) from the most upstream site. The pattern of changes in the number of taxa was related to changes in water chemistry, possibly associated with the influence of the calcareous Wenlock beds (see Chapter 1) on ionic concentrations in the R. Ithon. Thus, there were significantly more taxa collected at sites downstream from the confluence of the Wye with the Ithon where mean total dissolved

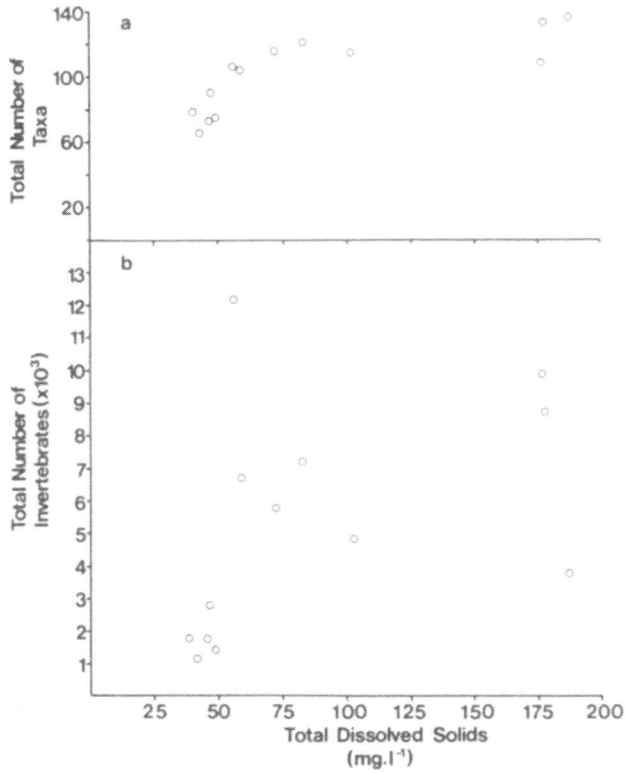

Fig. 19. Changes in a) total number of taxa and b) total number of invertebrates collected (September 1975–September 1977) in relation to total dissolved solids concentration.

solids (TDS) exceeded 50 mg/l, than at upstream sites where concentrations were lower (Fig. 19a).

Estimates of the density of macroinvertebrates at each site varied considerably over the period of study and, overall, ranged from about 500 to 22 000/m². As with the number of taxa, the mean density of macroinvertebrates collected at sites downstream of the Ithon was significantly greater than at upstream sites (Fig. 19b), and these data confirm the conclusions of Egglishaw & Morgan (1965) who have shown that invertebrate faunas in streams in Scotland are also restricted in base-poor upland waters compared with chemically richer lowland areas.

In general, the benthic faunas of upland sites (W4–W24) were dominated numerically by plecopterans in spring with mayflies, caddis and dipterans (principally chironomids) being more abundant in the summer months. Sites lower in the catchment were more variable with worms, mayflies, caddis, beetles and dipterans (chironomids) all being relatively abundant at different sites and times (Table 15).

Table 15 Proportional representation (percentages of total numbers collected) of major taxonomic groups in the Wye catchment, 1975–77. The approximate positions of sampling sites on the Wye (W), Elan (E) & Irfon (Ir) are shown on Fig. 2.

	W4	W8	W13	W17	E6	W18	W24	Ir1	W29	W35	W43	W53	W55	W61
Oligochaeta	8.5	16.4	26.4	15.9	18.0	19.1	4.4	10.8	8.9	6.8	22.2	17.6	24.8	13.1
Plecoptera	23.8	14.2	18.2	21.4	19.4	13.8	7.8	9.6	4.1	9.0	6.4	0.7	0.1	0.4
Ephemeroptera	17.6	26.2	9.2	19.7	7.5	18.6	18.8	16.4	28.9	18.0	25.5	13.0	2.9	14.1
Trichoptera	1.9	10.0	11.7	3.2	2.9	16.4	17.7	21.1	23.5	28.3	4.1	18.4	4.4	8.5
Coleóptera	3.3	7.4	15.2	11.2	0.6	6.4	6.6	11.7	11.1	10.0	11.8	13.8	10.5	20.2
Diptera (Chironomidae)	16.8	12.1	11.0	17.9	29.1	17.4	16.4	15.5	19.5	17.0	20.6	16.6	40.1	27.6
Diptera (Simuliidae)	23.6	3.7	1.9	2.1	18.1	2.2	26.9	4.3	0.2	3.6	0.2	9.3	0.1	1.1
Mollusca	N.R.	0.3	0.2	2.7	N.R.	4.1	0.2	0.6	1.6	6.7	0.9	2.0	2.7	6.3
Others	4.6	9.7	6.2	5.9	4.4	2.0	1.2	10.1	2.1	0.8	8.3	8.5	14.4	8.7

N.R. = Not recorded.

The impounded R. Elan (E6, Table 15), however, supported lower proportions of beetles, caddis larvae and mayfly nymphs than nearby sites on the R. Wye, and an intensive investigation of the Elan by Paull (1978) indicated that the development of organically rich deposits, containing substantial quantities of iron and manganese, was probably responsible for differences between the rivers. Much of the iron and manganese in the Elan originates from the upstream impoundment and from a water treatment works effluent, and the very stable compensation flow during the summer (see Fig. 3) favours the settlement of such material on the river-bed.

It is convenient to consider some of the numerically important invertebrate groups in the Wye separately and in detail. Some groups were inadequately sampled and receive no further consideration; others, like the Megaloptera, Odonata (adults of *Agrion splendens* (Harris) were collected from the lower Wye, but the nymphs, probably associated with *Ranunculus* plants, were never found), and some of the Hemiptera, which are not characteristic of fast-flowing rivers like the Wye, are not considered in detail; others which were not identified precisely and therefore present a ubiquitous distribution (e.g. Hydracarina), are also omitted.

Platyhelminthes

Only representatives of the Tricladida were recorded from the R. Wye and of these only *Planaria torva* was found at densities exceeding 1000/m². Some species showed restricted distributions. *Phagocata vitta,* which is a species characteristic of high, peaty ground and is known to occur in springs and underground streams, was found only in the upper reaches of the river where it was the principal triclad: *Dugesia lugubris* and *Dendrocoelum lacteum,* species generally recorded in productive lowland water bodies, were restricted to downstream reaches of the Wye (Fig. 20). The occurrence of *Polycelis felina* in the lower Wye seems unusual for a species characteristic of small streams.

Oligochaeta

The worms in the Wye catchment were represented by the naidids, enchytraeids, tubificids and lumbriculids and one species of lumbricid (*Eiseniella tetraedra*). The lumbriculids, *Lumbriculus variegatus* and *Stylodrilus heringianus,* were, overall, the most abundant species and occurred throughout the catchment. This ubiquitous distribution probably reflects the excellent water quality of the Wye, *S. heringianus,* in particular, generally being regarded as intolerant of organic pollution.

Most naidid species were widely distributed, but *Stylaria lacustris,* a species characteristic of still waters, and recorded at densities up to 4000/m² in the Wye during low flows in July 1976, *Pristina idrensis* and *Uncinais uncinata* were collected only in the lower reaches (Fig. 20).

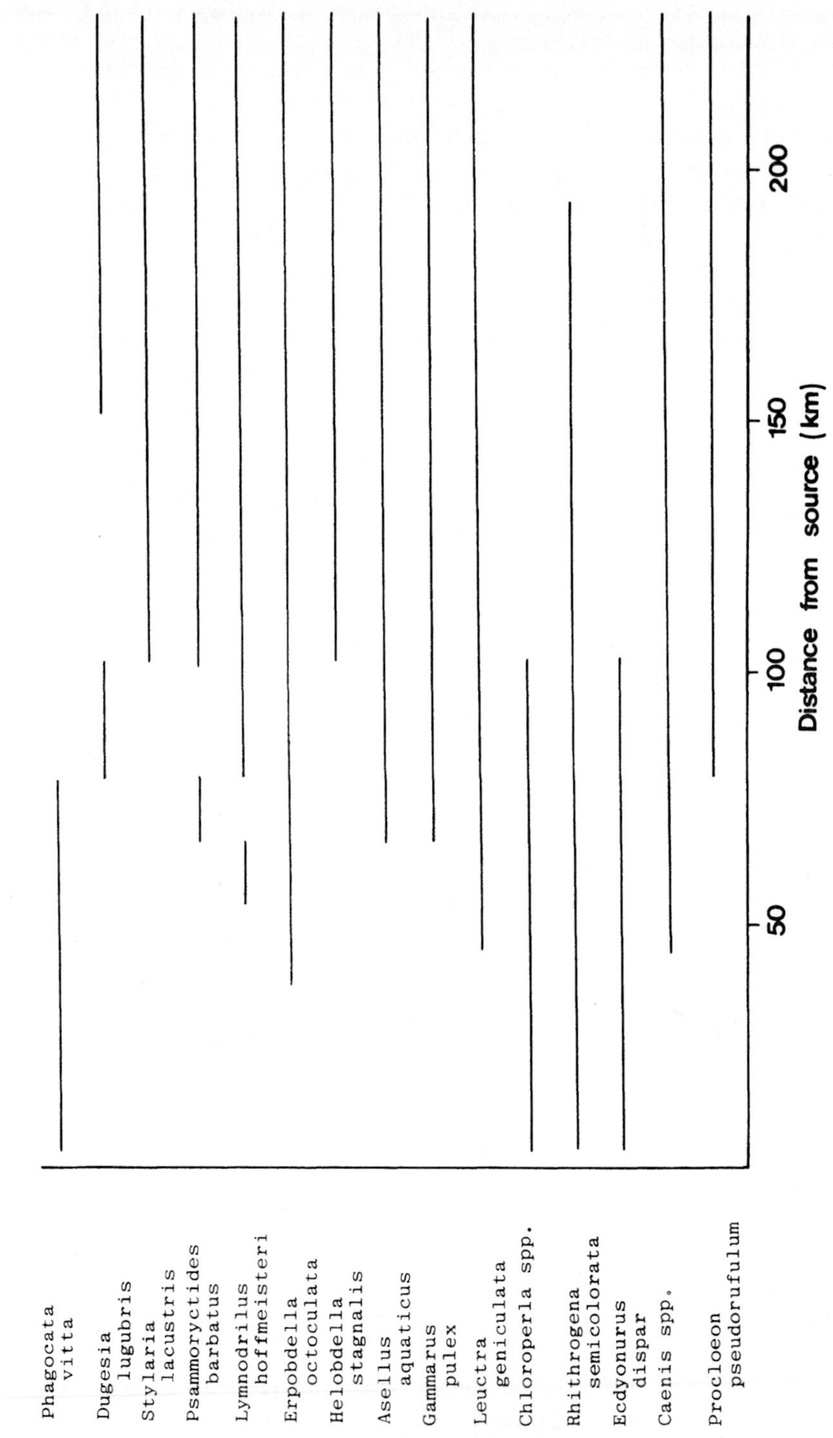

Phagocata
vitta

Dugesia
lugubris

Stylaria
lacustris

Psammoryctides
barbatus

Lymnodrilus
hoffmeisteri

Erpobdella
octoculata

Helobdella
stagnalis

Asellus
aquaticus

Gammarus
pulex

Leuctra
geniculata

Chloroperla spp.

Rhithrogena
semicolorata

Ecdyonurus
dispar

Caenis spp.

Procloeon
pseudorufulum

50   100   150   200

Distance from source (km)

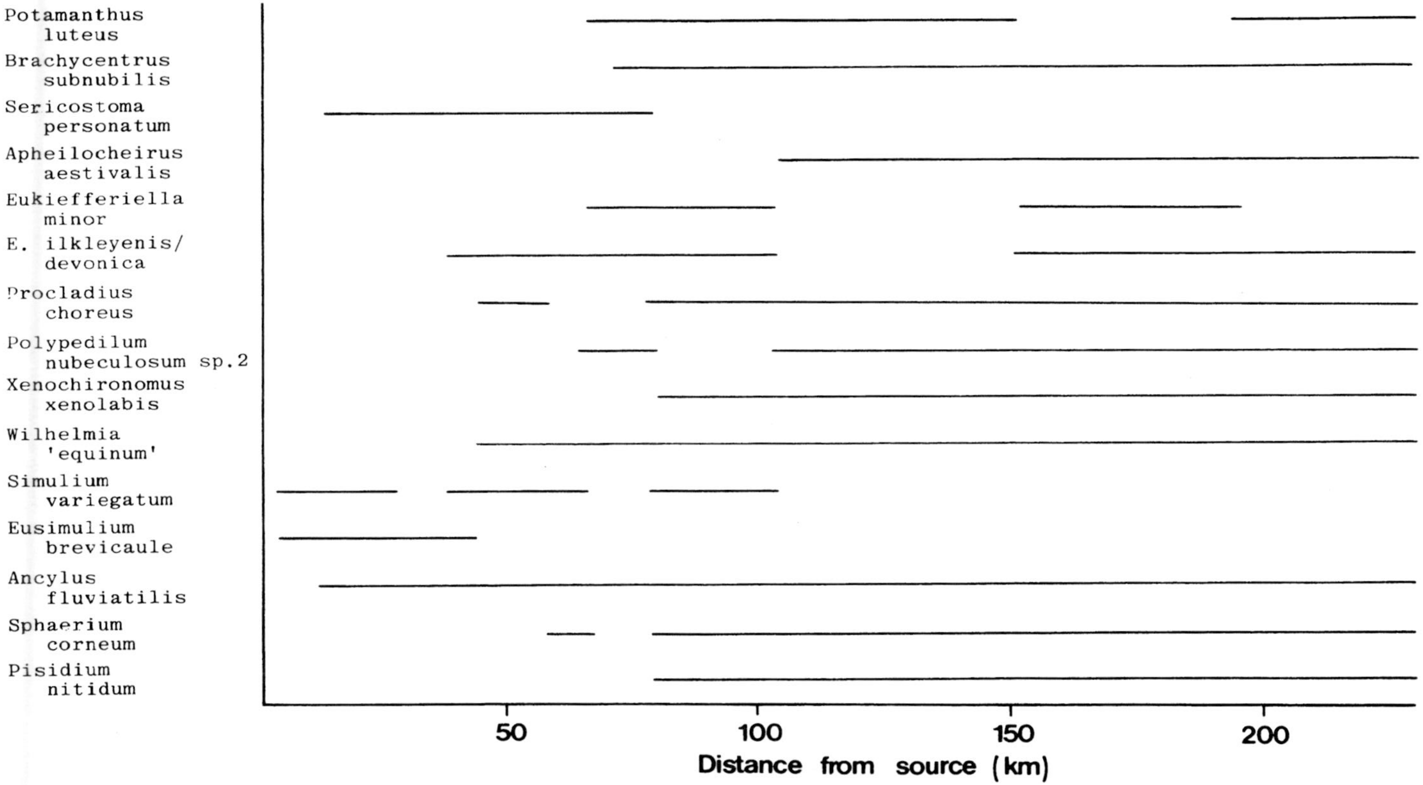

Fig. 20 Distributional records of selected macroinvertebrates in the main R. Wye (1975–77).

59

Representatives of the Tubificidae were generally found only in the middle and lower reaches, *e.g. Psammoryctides barbatus, Limnodrilus* spp. (Fig. 20), but *Peloscolex ferox* was found at most sites.

Hirudinea

All leeches were restricted in their occurrence, and none was collected upstream of the Wye–Elan confluence (W17) on the main river. *Glossiphonia complanata* and *Erpobdella octoculata* are often found in relatively soft waters, and this tolerance is reflected in the upstream penetration of these species compared with *Helobdella stagnalis,* which is more characteristic of hard, slow-flowing rivers (Fig. 20). *E. octoculata* was present in the R. Elan at a mean calcium concentration of 1.8 mg/l whilst *H. stagnalis* was only recorded at sites where the mean calcium concentration exceeded about 18 mg/l.

Crustacea

Asellus aquaticus and *Gammarus pulex* were found only at sites with calcium concentrations greater than about 9 mg/l (Fig. 20), but densities did not exceed 400/m². However, factors other than calcium concentration are also likely to play some part in their distribution: in the Endrick, Maitland (1966) found *Asellus* only in the most downstream 15 km of the river whilst *Gammarus* was more widely distributed, being absent only from the extreme upper reaches where calcium concentrations were less than 2 mg/l. Sutcliffe and Carrick (1973b), who studied upland streams in the Lake District, concluded that spatial and seasonal differences in pH rather than calcium concentrations explained the distribution of many insects and *Ancylus* and that pH was probably an important factor in governing the presence of *Gammarus,* the genus not being found at pH < 5.7. In the Wye *Gammarus* was restricted to sites with pH greater than about 7.

The oldest records of crayfish in the Wye catchment date from the early nineteenth century when it was reported that stocking of the Irfon had occurred: the species introduced was referred to as *Astacus fluviatilis* but was probably *Austropotamobius pallipes,* the indigenous species of this country. Hellawell (1971c) reported some incidental records of the species, but more recently Lilley et al. (1979) undertook a detailed survey of the crayfish of the upper Wye and the R. Lugg catchments.

The survey procedure – turning stones whilst moving upstream – was limited to depths of one metre and, therefore, in larger rivers to marginal areas, but revisited sites usually confirmed earlier records so the distributional data can be accepted as generally reliable. Trapping techniques, which are known to work well in lakes and some rivers like the Great Ouse in Bedfordshire, were also used but were not successful.

Table 16 Range of water quality in which *A. pallipes* was recorded. Concentrations in mg/l except where noted.

Calcium	6.1– 66.8
Sodium	7.0– 12.1
Potassium	0.8– 3.9
Magnesium	1.8– 12.3
Total hardness (as CaCO₃)	22.6–207.8
Conductivity (uS/cm)	60.0–390.0
pH	6.8– 8.5
Oxygen	7.0– 11.6

103 sites on 57 streams were visited, and more than 300 specimens of *A. pallipes* were examined. As in other studies, there were fewer males than females (0.6♂:1.0♀), and porcelain disease, caused by *Thelohania contejeani,* was widepread, affecting at least 3% of the population. *A. pallipes* was recorded at sites with a considerable range of water quality (Table 16), but its occurrence was randomly distributed with respect to all determinands except calcium, no specimens being found where the concentration was below 6 mg/l. It is known that crayfish have a high physiological requirement for calcium and that the uptake mechanism for absorption over the gills is not saturated until external concentrations reach about 16 mg/l: in fact, below this concentration there was a statistically significant predominance of absences at sites investigated in the Wye catchment, suggesting that calcium concentrations are important in limiting the distribution of the species. However, other factors, e.g. substrate, current velocity, local pollution, are also relevant in determining occurrence, and this was reflected in absences at some sites with high calcium concentrations.

Plecoptera

Stonefly nymphs formed a high proportion of the fauna throughout the Wye catchment, particularly during the spring (Table 15). Although many species are absent during the summer months *Leuctra fusca* nymphs hatch in April/May and develop during the summer, dominating benthic collections of stoneflies at this time: adults were recorded on the wing in August/September (Morris 1981).

Except for *L. fusca,* which was recorded at all sites, and *L. geniculata,* which was restricted to the middle and lower reaches, most stoneflies, e.g. *Chloroperla* spp., were found only in the upper and middle reaches of the Wye. (Fig. 20). The two species of *Chloroperla* in the Wye had similar distributions – this contrasts with their distributions in other rivers where *C. tripunctata* is much more restricted in its range than *C. torrentium.*

Incidental collections of adults at Monmouth in the lower Wye revealed the presence of *Brachyptera putata,* a rare species in the U.K., but one previously recorded from this location.

Ephemeroptera

In contrast with the stonefly nymphs, those of the mayflies are most abundant in summer collections, and 20 species have been identified in their juvenile stages. The nymphs and adults of the Ephemeroptera, or their mimics, are probably the game angler's favourite lure and some of the most notable of these, *Ephemerella ignita, Baetis scambus, B. rhodani, Caenis* spp. are found throughout the catchment with densities of nymphs often exceeding $1000/m^2$.

Nymphs of *Rhithrogena semicolorata* were also recorded at densities up to $1000/m^2$ and were found at most sites on the main R. Wye except in the most downstream reaches (Fig. 20). However, the species was not collected from the impounded R. Elan, where the absence of other ecdyonurids was noted. Indeed, Paull (1978) did not record any mayfly nymphs in the Elan during a summer survey, when *B. rhodani* and *E. ignita* were abundant in the Wye, and attributed this principally to substantial deposits of iron- and manganese-rich organic material of the river bed.

Several mayfly species showed restricted distributions in the Wye (Fig. 20), particularly *Ecdyonurus dispar* (W4–W35), *Caenis* spp. (W18–W61), *Baetis niger* (W29–W61), *Procloeon pseudorufulum* (W43–W61). *Ephemera danica* (W35–W61) and *Potamanthus luteus* (W29–W61). The distributions of some are probably related to current speed, *E. dispar* being associated with fast- and *P. pseudorufulum* with slow-flowing rivers; that of *B. niger* is generally associated with the presence of submerged macrophytes. *Potamanthus luteus* is rare in the U.K. and has been previously recorded only from the rivers Thames, Usk and Wye. Recent studies (Brooker & Morris 1980b) have not revealed the presence of *P. luteus* in lowland tributaries of the Wye, but a single specimen was collected from the nearby R. Usk in 1979. Its occurrence at only low densities ($\leqslant 50/m^2$) may restrict positive records of this species. Adults of *P. luteus* have been recorded at Hereford during July, and it seems likely from the size and temporal distribution of the nymphs that the species is univoltine with overwintering nymphs.

Two other rare species, *Ephemera lineata* and *Brachycercus harrisella,* have been recorded previously from the Wye but were not found in the surveys of Brooker & Morris (1980a). Nymphs of the latter have been recorded recently for the first time in the R. Towy in West Wales from reaches, not routinely sampled, with a silty substrate – this may explain the low incidence of records, or apparent absence, in the Wye and other rivers. Other unpublished records from North-East Wales and several rivers in Yorkshire suggest that it is probably a widespread, but rarely collected. species.

Trichoptera

Although more than 20 species of caddis larvae were found in the Wye – and some belonging to the families Beraeidae, Hydroptilidae and Leptoceridae were not

62

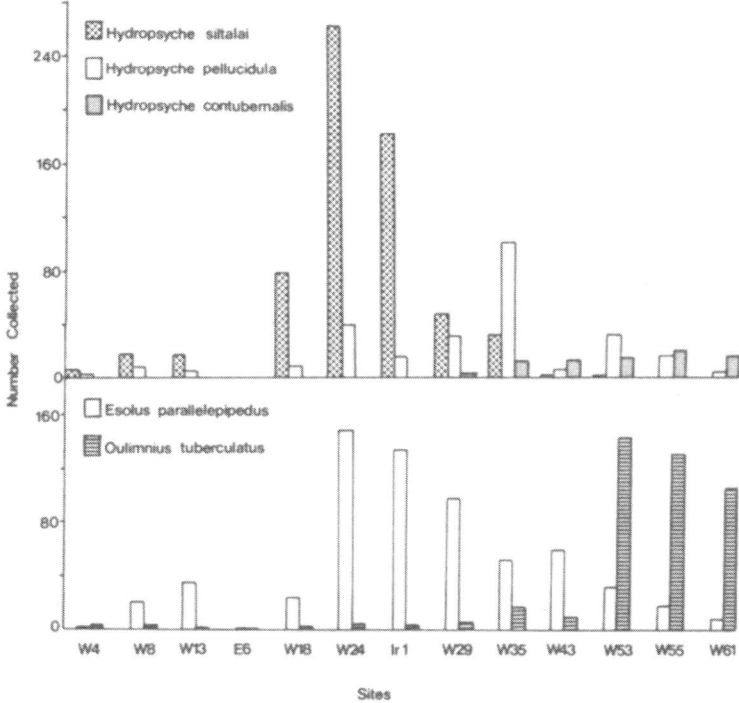

Fig. 21 The distribution and abundance of a) *Hydropsyche* spp. (Trichoptera) and b) Elminthidae (Coleoptera) in the Wye catchment (1975–77).

identified – the group was numerically dominated by three species of Hydropsychidae: *Hydropsyche siltalai, H. pellucidula* and *Cheumatopsyche lepida.* These species are also abundant in the nearby R. Usk (Hildrew & Edington 1979). Only *H. siltalai* and *H. pellucidula* were recorded in the R.North Tyne in the north of England (Boon 1978).

In the main Wye *H. pellucidula* was recorded at all sites, with a maximum density of 900/m², but *H. siltalai* (maximum density > 5000/m²) was not recorded from the lowest site in the catchment (Fig. 21a). *H. contubernalis,* which did not exceed 350/m² was found only at the sites below Erwood (W29–W61), and *Cheumatopsyche lepida* was regularly collected from sites at, and downstream of, Builth Wells. Although the pattern of distribution was generally similar to the nearby R. Usk, (Hildrew & Edington 1979) some minor differences were apparent. In the Usk, *H. pellucidula* was not recorded in the upper reaches of the river catchment and *H. siltalai* was found at all sites sampled. In both rivers *H. contubernalis* and *C. lepida* were restricted to downstream reaches, but the latter species penetrated further upstream in the Wye. *Diplectrona felix* (Hydropsychidae) was characteristic of small headstreams in the Usk but was not recorded in the Wye – probably because of the omission of such sampling locations.

Although hydropsychids are often found in great numbers below lake outfalls, where they filter out particulate materials, densities of these trichopterans in the impounded Elan (average, $5/m^2$) were significantly lower than at nearby sites on the R. Wye (average, up to $350/m^2$). This distribution probably results from the substantial deposits, referred to earlier, on the river bed of the Elan.

Polycentropus flavomaculatus, a regular member of the fauna, was generally distributed in the Wye catchment, but *Brachycentrus subnubilis,* most abundant at Erwood (W29) and Glasbury (W35), and *Sericostoma personatum,* characteristic of fast-flowing streams and rivers, were restricted to the middle/lower and upper reaches respectively (Fig. 20). Adults of *B. subnubilis* are often seen in swarms over the R. Wye, particularly in April.

Coleoptera

Beetles, particularly those belonging to the Elminthidae, occurred throughout the catchment and similar distributions are typical of other rivers in the U.K. There were four principal species: *Esolus parallelepipedus* and *Oulimnius tuberculatus,* which had maximum densities (adults and larvae) of about $4000/m^2$, and *Elmis aenea* and *Limnius volckmari,* with maximum densities of about $400/m^2$. Highest densities of *E. parallelepipedus* were recorded in the middle reaches (W24–W29), and those of *O. tuberculatus* were found in the lower reaches (W53–W61) (Fig. 21b). Little is known of the ecology of these species, and the reasons for the differences in distribution are not understood.

As with the hydropsychids, few beetles were found in the Elan. This is in contrast to the R. Tees, downstream of Cow Green Reservoir, in the north of England where increases in the numbers of beetles were recorded subsequent to the closure of the dam. Probably the iron- and manganese-rich deposits in the Elan inhibit the survival of beetles. This seems to be supported by the fact that the most abundant species in this river, *Limnius volckmari,* has been reported to be the only coleopteran surviving pollution from ferric hydroxide deposits in the Taff Bargoed in South Wales (Scullion & Edwards 1980).

Hemiptera

The only species regularly collected in surveys of the Wye was *Apheilocheirus aestivalis* which is characteristically found in the deeper reaches of fast-flowing rivers (Fig. 20). It was generally regarded as a southern species in the U.K. until its discovery in the R. Tweed in Scotland: *A. aestivalis* has also been recorded from the R. Dee in North Wales and from the Usk in the adjoining catchment to the Wye. These are amongst the largest rivers in Wales, exceeding 100 km in length and having substantial catchment areas ($>$ about 1000 km^2).

64

Diptera

The Diptera were represented principally by the Chironomidae and Simuliidae, the former representing up to 40% of the total number of macroinvertebrates collected.

Of the chironomids, the orthocladiine genera *Eukiefferiella* and *Cricotopus*, which had maximum densities of about $1500/m^2$, were generally the most abundant and were widespread throughout the catchment, *E. clypeata* and *E. discoloripes/verralli* the most common representatives, being recorded at all sites. *E. minor* and *E. ilkleyensis/devonica* were generally restricted to the middle and lower catchment (Fig. 20). Some abundant representatives of the Tanypodinae (*Procladius choreus*, up to $900/m^2$) and Chironomini *(Pohypedilum nubeculosum*, up to $7000/m^2$) were also collected from the middle and lower reaches of the main river (Fig. 20). *Xenochironomus xenolabis*, which is generally associated with sponges, was collected from sites in the middle and lower reaches.

Simuliidae (blackfly) larvae were particularly abundant on occasions at some sites in the upper Wye (W4 and W24) and the R. Elan (E6). The principal taxa were *Wilhelmia 'eqinum'*, which was restricted to the middle and lower reaches, and *Simulium reptans*, which was widely distributed. The 1970 survey undertaken by Nature Conservancy Council (Ratcliffe 1977) reported that *S. reptans* and *S. variegatum* were the predominant species in upland reaches of the Wye and surveys in 1975–77 confirmed that the latter species, together with *S. monticola* and *Eusimulium brevicaule*, were restricted to sites upstream of Hereford (Fig. 20).

There have been no reports from the Wye catchment of blackfly adults biting man or domestic animals, although *W. equinum*, *S. ornatum* and *S. reptans*, which are known biters, occur in the catchment. *S. austeni*, which has caused severe problems by biting people in Southern England (Hansford & Ladle 1979), was not recorded in the Wye.

Mollusca

The most widely distributed snail in the Wye catchment was *Ancylus fluviatilis*, recorded at all sites except the impounded R. Elan and the most upstream site sampled on the Wye (W4), these latter sites having mean calcium concentrations less than 3 mg/l. Other molluscs were much more restricted and were not found at sites with average calcium concentrations below about 9 mg/l, e.g. *Sphaerium, Pisidium* (Fig. 20). Some species were found only in the most downstream reaches of the river, particularly notable being *Theodoxus fluviatilis* (W53–W61) and *Planorbis albus* (W55 W61). Maitland (1966) found only three species of snail in the R. Endrick at calcium concentrations less than 10 mg/l – *A. fluviatilis, Lymnaea peregra,* (which was restricted to sites in the Wye with calcium concentrations >9 mg/l), and *Pisidium casertanum,* which was not recorded in the Wye.

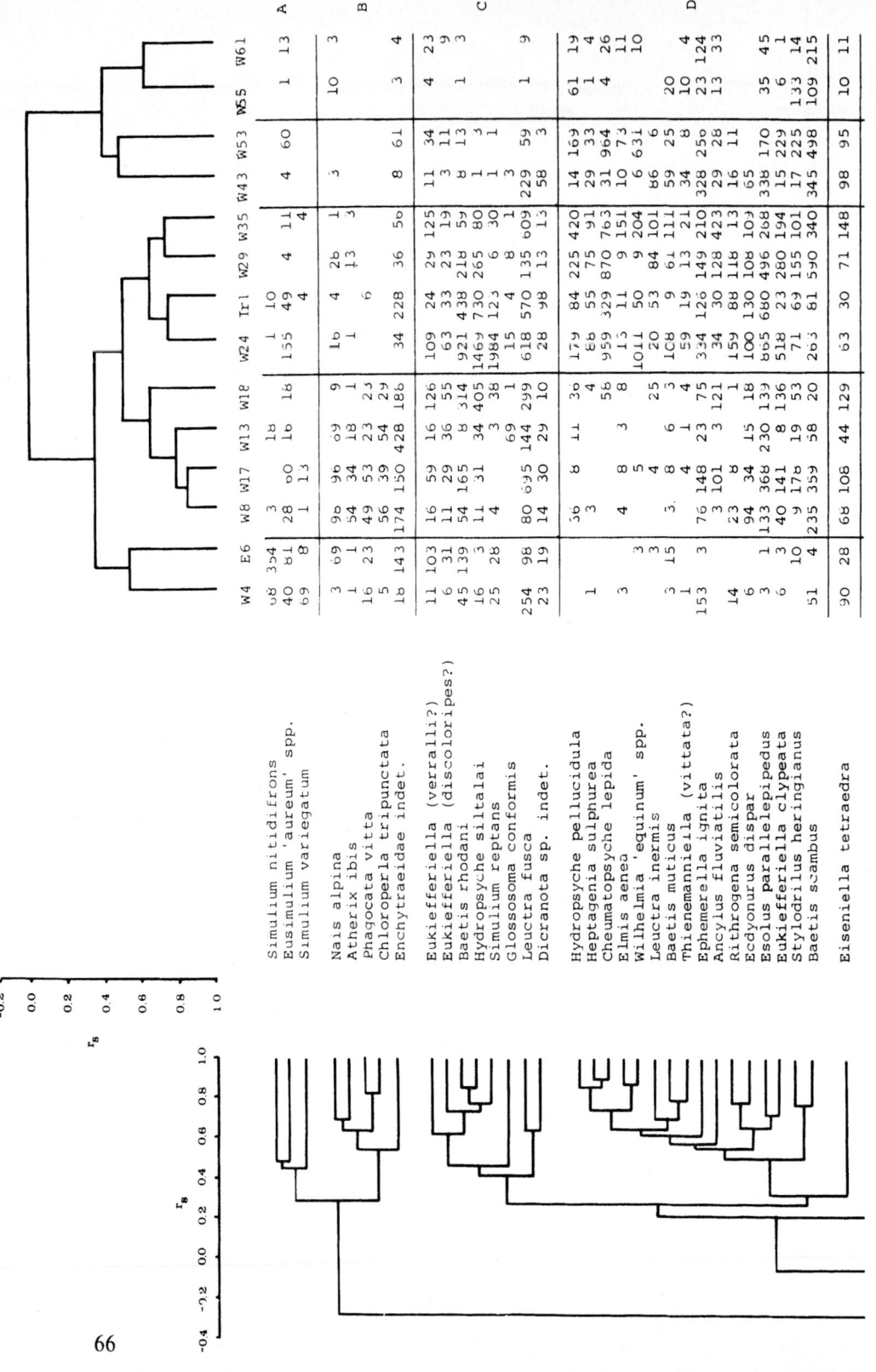

	W4	E6	W8	W17	W13	W18	W24	Irl	W29	W35	W43	W53	W55	W61
Simulium nitidifrons	68	354	3		18	18	1	10					1	13
Eusimulium 'aureum' spp.	40	81	28	60	16	16	155	49	4	11	4	60		3
Simulium variegatum	69	8	1	13			4	4	4	4				4

A

	W4	E6	W8	W17	W13	W18	W24	Irl	W29	W35	W43	W53	W55	W61
Nais alpina	3	69	96	96	69	9	16	4	26	1	3		10	3
Atherix ibis			54	34	18	1	1		13	3				
Phagocata vitta	16	23	49	53	23	23		6						
Chloroperla tripunctata	5		56	39	54	29								
Enchytraeidae indet.	16	143	174	150	428	186	34	228	36	50	8	61	3	4

B

	W4	E6	W8	W17	W13	W18	W24	Irl	W29	W35	W43	W53	W55	W61
Eukiefferiella (verralli?)	11	103	16	59	16	126	109	24	29	125	11	34	4	23
Eukiefferiella (discoloripes?)	6	31	11	29	36	55	63	33	23	19	11	11	1	9
Baetis rhodani	45	139	54	165	8	314	921	438	218	59	8	13	1	3
Hydropsyche siltalai	16	3	11	31	34	405	1467	730	265	80	1	3		
Simulium reptans	25	28	4		3	38	1984	123	6	30	3	1		
Glossosoma conformis				69			15	4	8	1				
Leuctra fusca	254	98	80	695	144	299	618	570	135	609	229	59	1	9
Dicranota sp. indet.	23	19	14	30	29	10	28	98	13	13	58	3		

C

	W4	E6	W8	W17	W13	W18	W24	Irl	W29	W35	W43	W53	W55	W61
Hydropsyche pellucidula	1		36	8	11	36	179	84	225	420	14	169	61	19
Heptagenia sulphurea			3		4	4	66	55	75	91	29	33	1	4
Cheumatopsyche lepida	3		4	8	3	58	959	329	870	763	31	964	4	26
Elmis aenea				5		8	15	11	9	151	10	73		11
Wilhelmia 'equinum' spp.		3	4	4		25	1011	50	9	204	6	631		10
Leuctra inermis	3	3		8	6		20	53	84	101	86	6	20	
Baetis muticus	3	45	3	4	1	3	108	9	61	111	59	25	10	4
Thienemanniella (vittata?)	1		4	4			59	19	13	13	34	8	23	124
Ephemerella ignita	153	3	76	148	23	75	334	126	149	210	328	250	13	33
Ancylus fluviatilis			3	101	3	121	34	30	128	423	29	28		
Rithrogena semicolorata	14		23	8	15	18	159	88	118	118	16	11		
Ecdyonurus dispar	6		94	34	230	139	100	130	108	103	65		35	45
Esolus parallelepipedus	3	1	133	368	8	136	865	680	496	268	338	170	6	1
Eukiefferiella clypeata	6	3	40	141	19	53	518	23	280	194	15	229	133	14
Stylodrilus heringianus		10	9	176	58	20	71	69	155	101	17	225	109	215
Baetis scambus	51	4	235	359	58	20	263	81	590	340	345	498	109	215

D

	W4	E6	W8	W17	W13	W18	W24	Irl	W29	W35	W43	W53	W55	W61
Eiseniella tetraedra	90	28	68	108	44	129	63	30	71	148	98	95	10	11

66

Fig. 22 — Distribution of the abundance of taxa groups in relation to site groups (1976–77, pooled data).

Dendrogram correlation scale: −0.4 −0.2 0.0 0.2 0.4 0.6 0.8 1.0

Cluster groups: **E**, **F**

Taxon	W4	E6	W8	W17	W13	W18	W24	Ir1	W29	W35	W43	W53	W55	W61
Polycentropus flavomaculatus	1	39	30	44	4	11	31	49	63	1	3	5	1	11
Hydracarina		1	28	73	29	13	50	340	55	4	8		16	16
Thienemannimyia spp.	19	84	20	143	16	18	124	65	66	38	83	26	40	36
Tanytarsus sp indet.	61	61	14	41	5	3	226	230	83	31	168	9	86	20
Brachycentrus subnubilus							3		134	495	29		15	106
Cricotopus bicinctus			9	1	1		8	1	24	16	5	25	14	94
Cricotopus sp. 3			1	1		4	6	18	29	39		4	89	18
Isocladius 'sylvestris' grp.		4	6			15	401	115	216	41			305	18
Orthocladiinae sp. 3	1		4				31	13	8	83	6		25	30
Orthocladiinae sp. 5	19						10	4	6	51	1	4	25	18
Microcricotopus rectinervis		1	3				4	3	3			1	5	73
Caenis moesta					6	33	18	75	608	50	348	134	19	100
Limnius volckmari	39	10	13	21	6	33	5	26	114	159	195	201	25	30
Asellus aquaticus							69	69	1	1	19	19	36	19
Polypedilus 'nubeculosum' sp. 2			1				21	26	1		128	40	1893	14
Rhyacodrilus coccineus		4	3	3		8	176	71	105	5	138	65	125	20
Polypedilum 'nubeculosum' sp. 1				13				69	174	41	120	461	523	63
Planaria torva						3				1	160	3	1173	31
Procladius choreus			1				10	6			13	44	231	36
Stylaria lacustris			10				31			1		4	1049	54
Aulodrilus pluriseta											26	4	8	204
Aphelocheirus aestivalis	3								3	26	4		8	51
Oulimnius tuberculatus			1	9		8	1	5	10	99	59	709	610	708
Synorthocladius semivirens			10	78		3	31	28	36	78	198	209	140	214
Limnodrilus hoffmeisteri									10	14	228	4	25	26
Hydropsyche contubernalis							1	6	20	75	56	21	55	93
Sphaerium corneum										70	3	29	60	33
Psammoryctides barbatus								3	3		29	21	94	75
Lymnaea peregra	1	4	1	4		15	13	1		1	21	11	4	70
Rheocricotopus spp. indet.	1	4	4	19			1	4	8	1	15	135	68	184
Micropsectra spp. 5		58	19	10		1	1	1	3	3	61	9	46	34
Cricotopus sp. 1	11	71	10				4	3	23	49	44	20	93	101
Eukiefferiella sp. 9		3	1	1		11	15	15	1	31	5	271	31	39
Lumbriculus variegatus	10	139	44	35	24	206	148	129	111	155	340	1081	609	115
Hydridae indet.	1		8	18	9	1	29		5	114		21	56	9
Cricotopus (trifascia?)	6	153				115		191	50		241	60	260	9
Peloscolex ferox	11	3	11	6	3	69	118	76	9	13	3	110	119	34
Rheotanytarsus sp. 1	11	15	75	15	51	25	76	76	16	98	19		135	25
Simulium ornatum	58	3					40	16		13		10		3
Simulium reptans var. galeratum	11	14		25			16			3		114	15	15

Fig. 22 Distribution of the abundance of taxa groups in relation to site groups (1976–77, pooled data). Groups are derived from average linkage clustering of Spearman rank correlations.

Shells of *Anodonta,* although not taken by the routine survey method, were found at several sites, usually on the river margins after floods, in the lower reaches of the Wye and also in the lower reaches of the R. Ithon, which was substantially richer in dissolved minerals than other rivers in the upper catchment.

Community analysis

The distribution of individual species (Fig. 20) provides a simple characterisation of different reaches of the R. Wye. Further spatial characterisation of the river based on groupings or associations of macroinvertebrates was undertaken in an attempt to interpret better the complex biotic and physico-chemical interactions governing such distributions. However, such interpretation of data from the R. Wye was constrained by the several changes in water quality and physical characteristics and causal relationships could not be estabhished.

Rank correlation coefficients were computed for the abundance of pairs of species at the sampling sites, and, between pairs of sites with respect to the abundance of those species present. Based on these correlation coefficients, both sites and species were classified into groups having affinity, using clustering techniques (Orloci 1975). These classifications formed the basis of a nodal analysis which provided a means of identifying species assemblages associated with particular groups of sites (Brooker & Morris 1980a) (Figs. 22 and 23).

In some cases the distribution of species is probably related to changing physical conditions (e.g. *Hydropsyche siltali,* Group C; *H. pellucidula,* Group D; *H. contubernalis,* Group F) and in others to changing water quality, particularly calcium (e.g. *Sphaerium corneum,* Group F). Whilst biological inter-dependence between species within a group was identified in the R. Cynon, South Wales (Edwards et al. 1975) using this analytical approach, where *Ancylus* was associated with larvae of a species of *Eukiefferiella,* known to live in the mantle cavity of the mollusc, similar examples have not been revealed from an analysis of the Wye data.

In a separate study of invertebrates in five upland tributaries of the Wye, Watson (1979) established significant relationships between both the biomass and density of invertebrates and certain water quality determinands (calcium, magnesium and conductivity). However, these determinands were themselves inter-related, restricting any simple interpretation of the factors influencing such invertebrate distributions.

Special investigations

More detailed macroinvertebrate studies, generally carried out with very specific objectives, have recently been undertaken by several workers in the Wye catch-

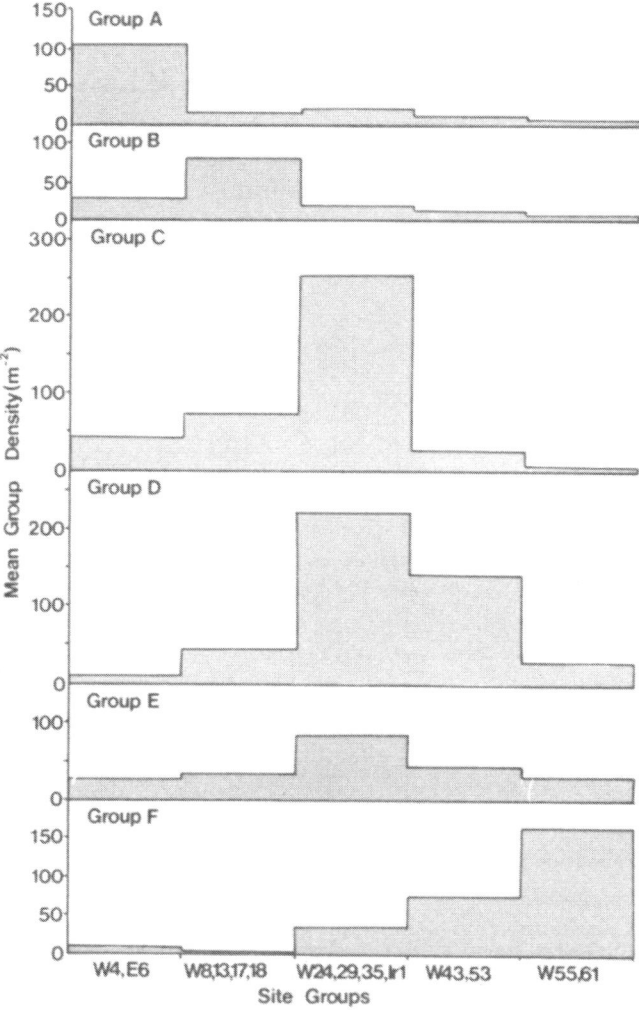

Fig. 23 Spatial distribution of taxa groups, A F (see Fig. 22) in the R. Wye 1976–77.

ment. The depth distribution of invertebrates within gravels, of particular importance in assessing the validity of conventional sampling methods, was examined in the upper catchment (Morris & Brooker 1979) and invertebrate drift was studied, principally in relation to seasonal and daily rhythms (Brooker & Hemsworth 1978; Hemsworth 1979; Hemsworth & Brooker 1979). Work on the emergence and flight patterns of aquatic insects, of considerable significance to salmonid feeding, was also carried out (Morris 1981). Another study considered the distribution of

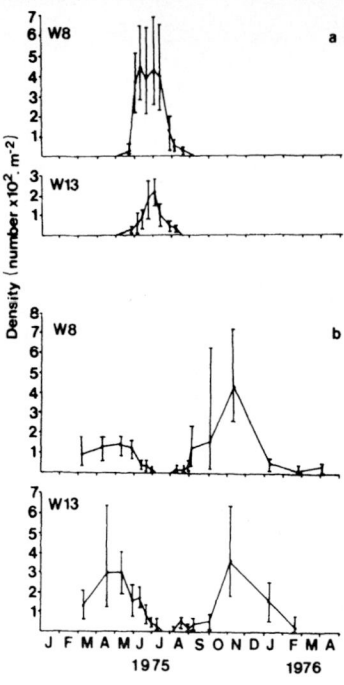

Fig. 24 Changes in the density of a) *Ephemerella ignita* and b) *Rhithrogena semicolorata* at two sites (W8 and W13) in the upper Wye. 95% confidence intervals shown.

zooplankton downstream from the Elan Valley reservoirs (Hopper 1978), and a programme of intensive sampling at certain sites in the upper catchment provided estimates of production of some of the more abundant species from size and abundance changes (Brooker & Morris 1978). The results of these investigations, where they contribute to our understanding of the ecology of the catchment, are described below.

Depth distribution of macroinvertebrates. Experiments, undertaken at one site only (W18), indicated that overall 59, 23 and 18% of benthic invertebrates were recorded in the top (0–11 cm), middle (12–22 cm) and bottom (23–33 cm) levels of basket samplers buried in the river substrates (Morris & Brooker 1979). Some organisms were generally confined to the top level, e.g. mayfly nymphs, blackfly (Simuliidae) larvae, snails, and could be satisfactorily characterised by surface samples, but others were either more uniformly distributed (worms, midge larvae) or increased with depth at certain times (e.g. *Sericostoma* larvae).

Life cycle descriptions and production estimates. The life cycles of four invertebrate species were described in detail and estimates of production undertaken (Brooker & Morris 1978; unpublished data):

a) *Ephemerella ignita*. As in most rivers in the U.K., peak densities of nymphs in the Wye (about 500/ m²) were recorded during June and July, the larvae only being recorded during the period May–August (Fig. 24). Peak summer densities in the R. Endrick (Scotland) and Bere stream (Southern England) were about 7000 and 1500/ m² respectively. Nymphs were recorded during the winter period in the Bere stream but not in the Endrick (Maitland 1965; Bass 1976) or Wye, probably reflecting the effect of temperature regime on egg development (Elliott 1978).

Production of *E. ignita* was estimated from changes in the mean individual weight and population density, making corrections for periods when recruitment and emergence influenced estimates of mean weight (Brooker & Morris 1978). Annual production varied from 164 to 794 mg dry wt/ m²/y with cohort turnover ratios of 4.5 and 6.6 (Fig. 25).

b) *Rhithrogena semicolorata*. The nymphs of *R. semicolorata* and *R. haarupi* are indistinguishable, but the development of nymphs in the upper reaches of the Wye suggested that all were *R. semicolorata*, and this is supported by collections of

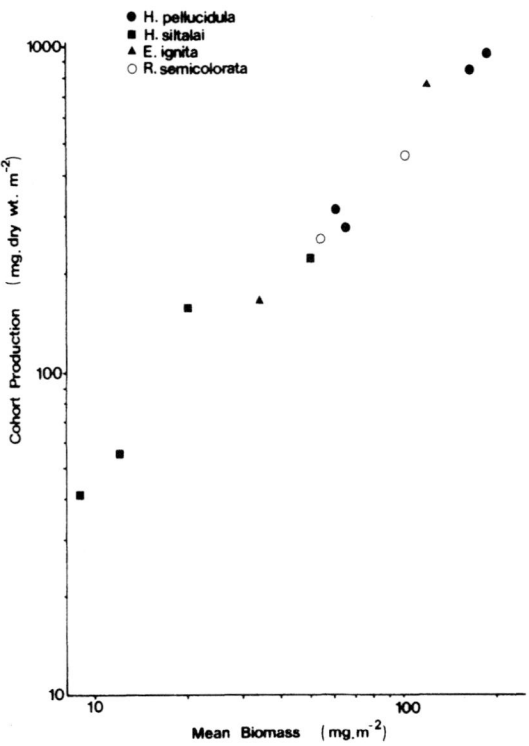

Fig. 25 Relationship between annual production (P, mg dry wt/ m²) and average cohort biomass (B, mg dry wt/ m²) of mayflies and caddis larvae in the upper Wye.

71

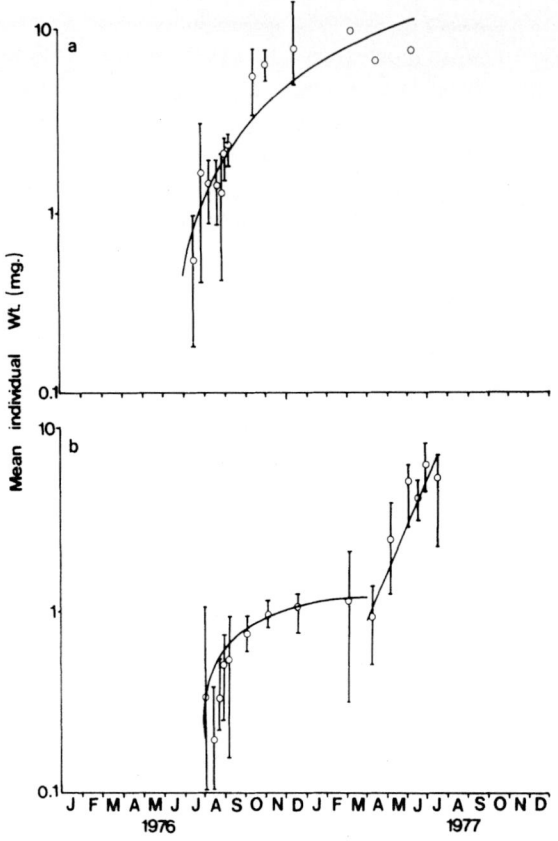

Fig. 26 Changes in the mean weight of the 1976 year class of (a) *H. pellucidula* and (b) *H. siltalai* at W18. 95% confidence limits shown except where sample size <3. Changes in weight estimated from geometric and exponential models shown as solid lines.

adults (Morris, 1981). The life cycle of *R. semicolorata* contrasted with that of *E. ignita,* with maximum densities (about 400/m²) being recorded during November–December, nymphs not being found during July. Similar life cycles have been recorded in the Welsh Dee and R. Endrick, but peak densities in these rivers (100/m²) were much lower (Badcock 1949; Maitland 1964). Annual production in the Wye was estimated to vary from 249 to 435 mg dry wt/m²/y with cohort turnover of about 4.5 (Fig. 25).

c & d) *Hydropsyche* spp. The life cycles of both *H. pellucidula* and *H. siltalai* in the Wye, interpreted from changes in larval density and individual weight and from adult flight patterns, took about a year, but there were differences in the timing and pattern of development (Fig. 26). These differences in life cycle have been recorded in other rivers in the U.K. (e.g. Boon 1979), and it has been postulated that they may act as an ecological isolation mechanism (Hildrew & Edington 1979). The estimates of annual production of the two species in the Wye were considerably different – *H. pellucidula,* 280–953 mg dry wt/m²; *H. siltalai,* 42–229 mg dry wt/m² – but cohort turnover was similar, generally ranging from about 4 to 5 (Fig. 25).

Drift of macroinvertebrates. Those organisms carried in suspension by water currents in rivers like the Wye are potentially important colonisers of new or denuded habitats and conversely the magnitude of their loss downstream may impair the viability of upstream populations. Since many of the studies on the Wye were related specifically to plans to regulate the flow of the river a special investigation of drifting macroinvertebrates was undertaken (Brooker & Hemsworth 1978; Hemsworth 1979; Hemsworth & Brooker 1979) in order to assess general loss rates and to describe the effects of artificial releases of reservoir water on the drift of benthic populations. Except for studies in the R. Tees (Armitage 1977, 1978) and R. Leven (Elliott & Corlett 1972) there are few studies in the U.K. of drift in rivers as large as the Wye, most investigations being restricted to small streams.

The seasonal and daily patterns of drifting macroinvertebrates at five sites were characterised. Daily variations resulted principally from the activities of nymphs of mayflies and stoneflies and blackfly larvae, maximum drift density occurring most frequently during darkness, the definition of peaks being influenced by sampling intensity (Fig. 27). However, on some occasions maxima were recorded immediately before or after sunrise and during the long nights of winter two peaks in density were sometimes recorded. Some organisms, e.g. *Hydra,* water mites, generally had highest drift densities during daylight (Fig. 27).

Estimates of mean daily density ranged from about 6×10^{-2} to $783 \times 10^{-2}/m^3$, with maxima generally during July and August. Total drift density was correlated with total benthic density. Such drift densities reflected total numbers drifting downstream of between 8 and $1400 \times 10^3/d$, but it was estimated that downstream population displacement of selected insect species over the aquatic phase of their life cycle was not likely to be greater than about 10 km (Hemsworth & Brooker 1979), and the normal dispersive flight of adults probably provides adequate compensation for downstream losses in the R. Wye.

Although the inverse relation between drift density and river flow, demonstrated in the R. Tees in Northern England, was not evident in the R. Wye, nevertheless the importance of changes in flow to drifting organisms was illustrated when the flow of the Wye was experimentally increased from about 2 to 5 m³/s following a summer release of water from the Elan Valley reservoirs. On the first day of the release the number of macroinvertebrates drifting at a site about 20 km downstream of the reservoirs was about seven times that of the preceding day: it decreased on the second day but was still three times greater. Some organisms, such as the larvae of the midge *Rheotanytarsus,* responded immediately to the increased flow whilst others, such as nymphs of the mayfly, *Ephemerella,* increased eight hours later, during the night – showing an enhanced response of their normal time of maximum abundance in the drift (Fig. 28).

73

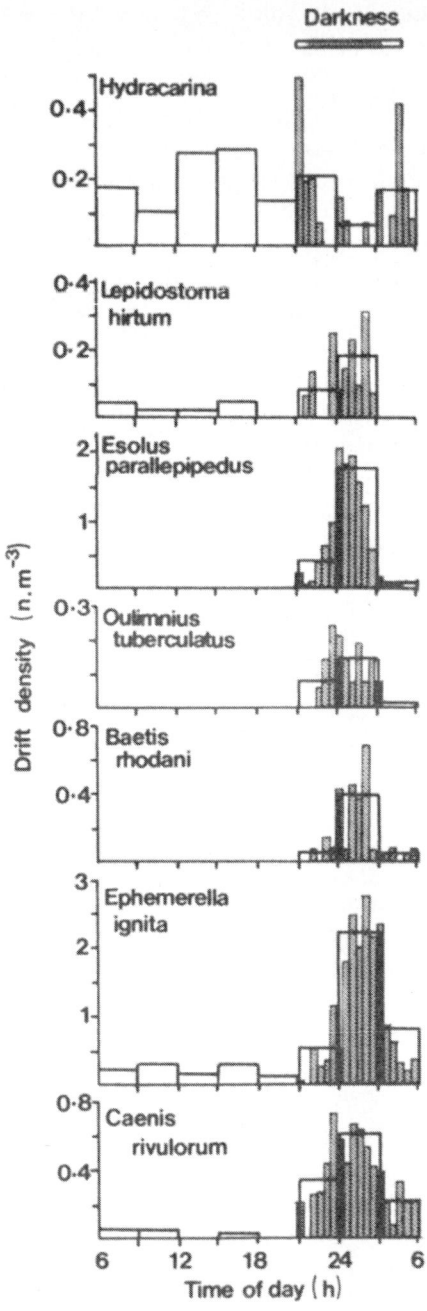

Fig. 27 Daily changes in drift density (n/m³) of selected taxa reflected by sampling intervals of 3 h and 0.5 h (upper Wye, June 1976).

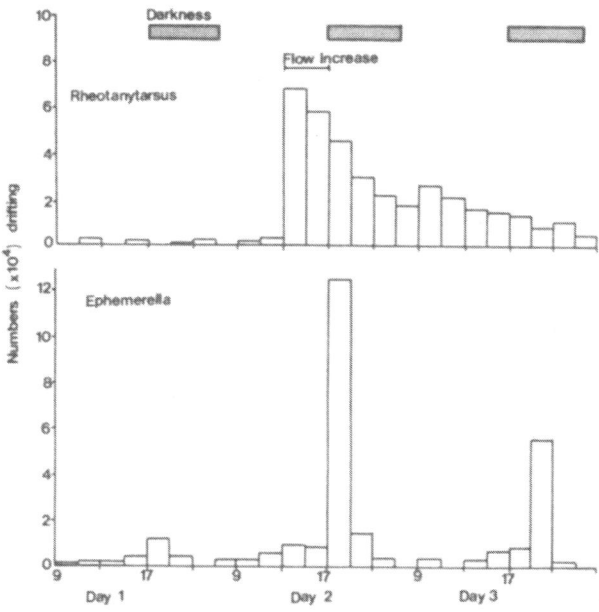

Fig. 28 Total numbers per 3 h of larvae of the midge *Rheotanytarsus*, and nymphs of the mayfly *Ephemerella ignita* before and during an increase in water flow resulting from a release of water from impoundments in the Elan Valley.

Adult insects in the upper Wye. During the period 1976–78, Morris (1981) studied the adult stages of aquatic insects at Newbridge-on-Wye, about 60 km from the source of the Wye. 128 adult taxa, most identified to species, were collected by a variety of methods including light-, suction- and emergence-traps (Appendix 4).

With all trapping techniques the Diptera (Chironomidae) contributed the greatest numbers both of taxa and individuals and, except for stoneflies collected in emergence traps, the caddis flies were the next most abundant group (Table 17). Such distributions are clearly related to the selective nature of the different trapping techniques, and, as might be expected, collections of adults emerging directly from the water surface (emergence traps) reflected most closely the description of the benthic juvenile insect fauna based upon samples of the river bed (see Table 15). Results from emergence traps suggested that the total number of aquatic insects emerging from the Wye at Newbridge was about 9000/m²/y. Estimates from Canadian streams are within the range 4500–21 000/m²/y (Judd 1953; Harper 1978), and lakes and reservoirs in the U.K. yield from about 6000 to 27 000/m²/y (Morgan & Waddell 1960). These emergence estimates for the Wye are, as is generally found, appreciably lower than numbers calculated from estimates of the densities of late-instar nymphs and larvae together with information of the voltinism of dominant species.

Table 17 Proportional representation (%) of adult taxa and numbers of insects in different traps at Newbridge-on-Wye over the period 1976-78.

	Taxa			Numbers		
	Light	Suction	Emergence	Light	Suction	Emergence
Plecoptera	4	2	16	<0.1	0.1	19
Ephemeroptera	4	2	7	<0.1	0.2	5
Trichoptera	29	20	16	7	3	8
Diptera						
–Chironomidae	59	73	58	93	97	67
–Simuliidae	4	–	3	<0.1	–	1
Megaloptera	–	2	–	–	0.1	–

Except for emergence traps, collecting methods reflected the flight activity of insects rather than their life-history patterns. For example, at Newbridge-on-Wye the number of aquatic insects collected in the light trap was closely related to air temperature. Obviously other partially related factors, such as humidity, wind and rainfall are also likely to influence flight activity. In addition, the fact that most caddis flies and many aquatic midges are nocturnal fliers accounts for their relatively high abundance in light traps: most stoneflies and mayflies fly by day. The seasonal distribution of adult flight activity, derived from all methods of capture, is shown in Appendix 4.

Morris (1981) also showed, using sticky traps, that the number of adult aquatic insects decreased rapidly with increasing distance from the river. In September the aerial density of adult insects 35 m away from the water's edge was only 7% of the density over the river, there being little difference between insect numbers at heights of one and two metres. In this series of samples the adult insects were chiefly chironomids: other workers (Roos 1957) have recorded adults of aquatic insect groups up to 5 km from the nearest potential emergence site.

Macroinvertebrates in the Elan Valley reservoirs

Zooplankton

The only published report of zooplankton in these reservoirs is that of Round (1956) who made incidental observations of the fauna when examining the phytoplankton (see Chapter 3). Crustaceans and rotifers were recorded during the period 1952-55 and the former, principally *Bosmina* and *Diaptomus,* were the most abundant zooplankters. Both genera showed seasonal changes in numbers with a single annual period of increase during late summer and early autumn.

Table 18 Species of crustacean recorded drifting downstream of the Elan Valley reservoirs.

Cladocera	Copepoda
Bosmina coregoni	
var. *obtusirostris* Sars	*Diaptomus gracilis* Sars
Holopedium gibberum Zaddach	*Cyclops (abyssorum/strenuus)*
Ceriodaphnia quadrangula (O. F. Muller)	
Chydorus sphaericus (O. F. Muller)	*Cyclops viridis* (Jurine)
Alonopsis elongata Sars	
Alona affinis Leydig	

Some 20 years later Hopper (1978) studied the drift of zooplankton from the Caban Coch Dam downstream in the rivers Elan and Wye during the period July–September 1977. Nine species of crustacean were recorded (Table 18). Samples collected from the body of the reservoir beside the dam in late July and early August indicated that *Bosmina coregoni* var. *obtusirostris* and *Holopedium gibberum* were the most abundant species with reported maximum densities of up to $449 \times 10^3/m^3$ *(H. gibberum).* Such high densities are generally restricted to more productive water bodies and therefore the high densities in these reservoirs deserve further investigation. The density of both of the principal species varied with depth, the major proportions being in the upper 10 m of the water column. *H. gibberum,* typical of soft water lakes, was particularly restricted to the upper layers by its buoyant, gelatinous body and in Caban Coch most of the population was recorded in the surface (<1 m) layers. In addition, horizontal distributions are clearly patchy, since total numbers in the sampled water column changed substantially over a daily cycle.

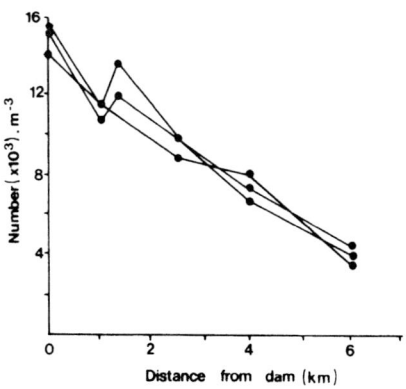

Fig. 29 Changes in the density of *Ceriodaphnia* in the R. Elan downstream of Caban Coch Reservoir on three occasions in July–August 1977.

In late August *Ceriodaphnia quadrangula* was the principal species in Caban Coch reaching a maximum density of $25.8 \times 10^3/m^3$ (depth 10 m) and the densities of *B. coregoni* (max., $10.4 \times 10^3/m^3$) and *H. gibberum* (max., $6.6 \times 10^3/m^3$) were much reduced.

The most abundant crustaceans in the water discharged from the Caban Coch Dam (drawn from a depth of about 25 m) during August 1977 were *B. coregoni* and *C. quadrangula,* and reported maximum densities immediately downstream of the reservoir were about 5 and $15 \times 10^3/m^3$ respectively. Of these less than 30% drifted to the confluence with the Wye, a distance of about 6 km (Fig. 29). Subsequently densities were generally substantially diluted by water draining from the upper Wye catchment, and few zooplankters were recorded further downstream than about 10 km below the Elan-Wye confluence. There appeared to be some differences in the rate of loss of different species from the water column. Ward (1975), who studied zooplankton loss below a reservoir in the U.S.A., has reported a negative relationship between the distance crustaceans drift downstream of an impoundment and species size, large species being more persistent than small.

Reported total densities of zooplankton ($4.7-31.7 \times 10^3/m^3$) were substantially greater than those reported in rivers downstream of Windermere, $1.7 \times 10^3/m^3$ (Elliott & Corlett 1972), in the Lake District and Cow Green Reservoir, $1.3-2.9 \times 10^3/m^3$ (Armitage & Capper 1976), in the north-east of England.

Overall the losses from the Elan Valley reservoirs were estimated to represent approximately 2 t dry wt/y with about 70% of the loss being retained, initially, within the lower Elan (about 6 km). Annual losses from Cow Green Reservoir were estimated to be about 0.15 t with only about 1–2% of this zooplankton drifting more than 6.5 km (Armitage & Capper 1976).

The immediate fate of such organisms in the Elan has not been studied but Armitage & Capper (1976) reported that stomachs of trout in the Tees below Cow Green Reservoir contained large quantities of micro-crustaceans. They are therefore eaten directly by fish, and it is probable that invertebrates will also feed directly upon them, although in the Elan this has not led to enrichment of the invertebrate fauna, possibly because of the predominant effects of the iron- and manganese-rich sediments in this river.

5. Fish

Introduction

During the twentieth century there have been three major scientific investigations of the fish of the catchment, in addition to the continuing routine management studies carried out by the statutory fishery agency, now the Welsh Water Authority. Mr J. A. Hutton, during the period 1912–47 published many papers, principally in the Salmon and Trout Magazine, and several books on aspects of salmon fishing and biology. His writings form the collection 'Salmon Scales', deposited in the library of the Freshwater Biological Association, Windermere. Dr J. M. Hellawell of Liverpool University studied the chub, dace, roach and grayling in the R. Lugg and R.Llynfi during the period 1964–66. In 1975 Dr A. S. Gee and Dr N. J. Milner with their colleagues of the University of Wales Institute of Science and Technology started a series of field investigations, mainly of salmon and trout in the upper Wye and its tributaries: this field-work was supplemented by analyses of catch statistics of salmon.

Some of these investigations were principally contributions to the basic ecology of the fish species concerned, and only those studies of specific relevance to fish stocks of the Wye will be included in this chapter. Furthermore, these investigators focussed on very different aspects of fish biology and, in consequence, the very different treatment given to each species in the section dealing with general biology reflects the idiosyncrasies of primary investigators and the period in which they worked rather than the quirks of authorship. This section will be preceded by a brief account of the distribution of species within the catchment.

Distribution

The Wye contains 29 fish species if one includes the lampreys, but this number continues to increase, the barbel, first recorded in the 1970s, probably being the most recently established (Table 19). Very occasionally species other than those in Table 19 have been caught in the Wye, the most impressive perhaps being the sturgeon, one specimen caught upstream of Hereford in 1846 and now in the Hereford Museum, weighing 82 kg.

Table 19 List of fish species recently recorded in rivers of the Wye catchment. See Maitland (1972) for nomenclature.

Lampetra planeri (Brook lamprey)	*Abramis brama* (Bream)
Lampetra fluviatilis (River lamprey)	*Alburnus alburnus* (Bleak)
Petromyzon marinus (Sea lamprey)	*Phoxinus phoxinus* (Minnow)
Alosa fallax (Twaite shad)	*Scardinius erythrophthalmus* (Rudd)
Alosa alosa (Allis shad)	*Rutilus rutilus* (Roach)
Salmo salar (Atlantic salmon)	*Leuciscus cephalus* (Chub)
Salmo trutta trutta (Sea trout)	*Leuciscus leuciscus* (Dace)
Salmo trutta fario (Brown trout)	*Noemacheilus barbatulus* (Stoneloach)
Salmo gairdneri (Rainbow trout)	*Anguilla anguilla* (Eel)
Thymallus thymallus (Grayling)	*Gasterosteus aculeatus* (3-spined stickleback)
Esox lucius (Pike)	*Perca fluviatilis* (Perch)
Cyprinus carpio (Carp)	*Gymnocephalus cernua* (Ruffe)
Barbus barbus (Barbel)	*Cottus gobio* (Bullhead)
Gobio gobio (Gudgeon)	*Platichthys flesus* (Flounder)
Tinca tinca (Tench)	

Although a systematic survey of the distribution of fish species within the catchment has not been undertaken, fishery bailiffs recently contributed towards an assessment of species distributions (Fig. 30), based principally on catches of anglers (Shaw 1977). Distributions of some species appear more restricted than in earlier reports (Page 1908), probably a reflection of the survey procedure. Several species, such as brown trout, salmon, bullhead, stoneloach, chub, dace, roach, eel, grayling, gudgeon, pike and minnow are widely distributed in the principal rivers of the catchment and in many tributaries, although chub, roach and pike are never found on the Wye upstream of Rhayader and dace only occasionally. Hellawell (1973) refers to the remarkable differences between the abundance of chub and dace in some tributaries and in one exploratory survey of the R. Irfon, which he carried out between Llanwrtyd Wells and Llangammarch Wells, he found many chub but no dace: in contrast, in the R. Ithon he found predominantly dace. However, such surveys merely highlight the patchiness of distributions and the unreliability of sampling techniques, for anglers' returns indicate the presence of considerable numbers of chub in the R. Ithon.

Other species, such as perch, carp and tench seem more localised in their distributions: the perch is rarely if ever caught on the Irfon and Ithon, the carp seems principally restricted to the Monnow and mid-Wye, and the tench, nowhere an abundant species, although similarly distributed, extends its range into the Lugg system. The bleak has recently increased in abundance and is now widely distributed in the lower catchment. The ruffe and rudd seem surprisingly localised in view of earlier more extensive distributions (Page 1908): the ruffe is principally restricted to the Monnow and Trothy and the rudd to mid-reaches of the Wye. The flounder has been recorded upstream of Hereford, 100 km from the Severn Estuary, and the shads are reputed to reach Newbridge-on-Wye, 190 km from the estuary, on rare occasions, although generally they are not found upstream of Builth Wells.

The rainbow trout, widely stocked in the Wye catchment, breeds occasionally, and fry from natural spawning have been found near Builth Wells (Gee, personal communication) and Ross-on-Wye (Staite, personal communication). Sea-trout, although regularly caught in the main river, are never abundant: they are particularly scarce in the Lugg, Trothy and Monnow – being prevented from ascending the Monnow by Osbaston Weir at Monmouth.

The sea-lamprey commonly spawns in the Ithon and in the Wye downstream of Builth Wells, and, although its distribution is fairly widespread in the remainder of the catchment, its comparative abundance is not well established. The lampern, or river-lamprey, spawns commonly in the lower Wye and Lugg system with some further spawning areas in the lower Irfon. The non-migratory brook-lamprey is widely distributed throughout the catchment.

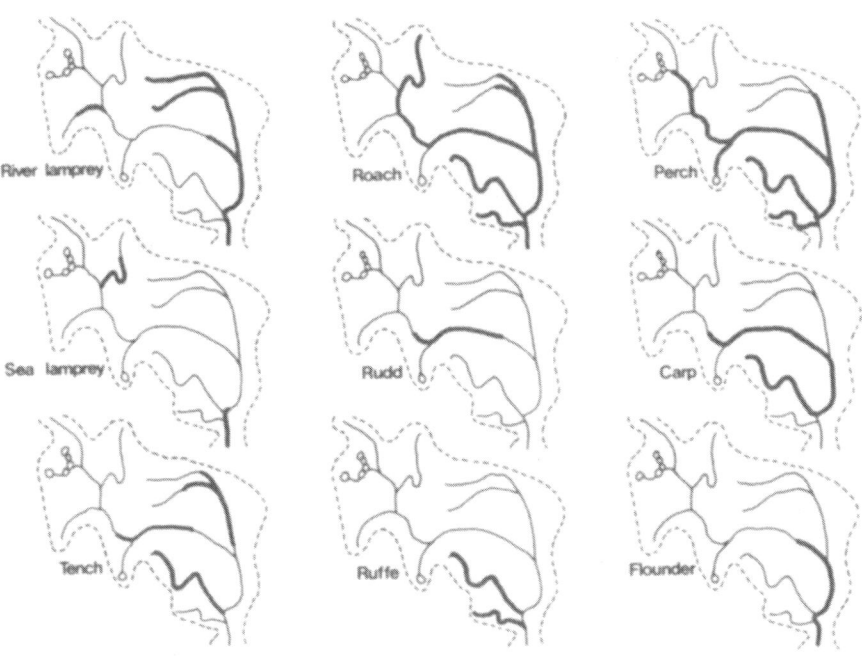

Fig. 30 Distribution of various fish species in the R. Wye and its main tributaries: reaches where such species are regularly found are shown by a broad line.

General biology

Shad

Although the twaite shad is fairly widespread and common in British rivers, predominantly in the west (Maitland 1972), the larger Allis shad is restricted to a few rivers on the west coast and in those it is rare. Shad were once more abundant and extensively distributed in Britain, equalling in value the salmon in the lower reaches of such rivers as the Severn where their decline in the eighteenth and nineteenth centuries has been attributed to the construction of navigation weirs (Day 1890). In Wales the Allis shad is regularly recorded in the Wye. Earlier writers (Ellison 1935) have suggested that it appears in freshwater in late April, males arriving first and in small shoals, with the larger females, weighing up to 1.6 kg, arriving later and generally singly. Twaite shad usually appear in the river early in May, and both species spawn in shallow water in late May and June, their spawning activity being associated with surface tail-thrashing of the water, apparently only by males. After spawning, survivors return to sea, soon to be followed by recently hatched fry. In most years substantial mortalities of adults occur in spawning areas. A suggestion by Tate-Regan (1911) that young overwinter in fresh-waters has recently been confirmed by Claridge & Gardner (1978) from fish collections on the screens of power stations in the Severn Estuary. It seems that some fry do not descend to the estuary during the summer and autumn but remain in their parent rivers until the following spring when the temperature rises above 9 °C: it is remarkable that these overwintering fish, which reach 6 cm in length, have not been reported in either the Severn or Wye. The studies of Claridge & Gardner (1978) confirm the rarity of the Allis shad for in three years of intensive sampling only two adults were caught.

Chub

The chub, considering the river as a whole, is possibly the most abundant coarse fish species and that most commonly caught by anglers. In several European rivers a preponderance of females has been recorded: in the Wye, Hellawell (1971a) found a high proportion of females only in the older age classes (>10y). For the chub populations sampled by Hellawell in the mid-1960s, the 1959 year class was dominant. The summer of 1959 was particularly sunny and dry, and it seems that dace and roach also produced large numbers of fry during that season which survived to dominate the populations for several years. Such dominant year classes are common with cyprinids.

Chub growth is remarkably similar in the upper and lower Wye (Gough, personal communication), in tributaries (Hellawell 1971a) and in rivers elsewhere in Southern Britain (Mann 1976; Cragg-Hine & Jones 1969). In investigations

82

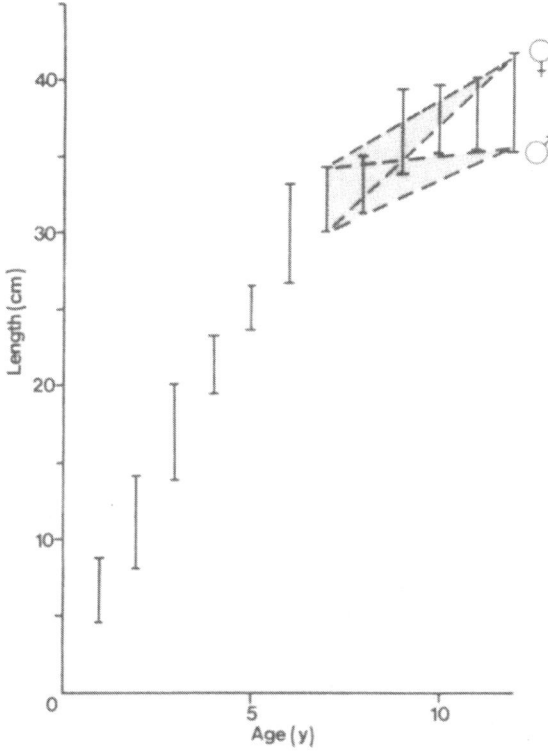

Fig. 31 Relationship between length and age of chub in several rivers in southern Britain, including the Wye and its tributaries.

where the growth of the sexes has been distinguished, it is evident that after six or seven years females grow faster than males (Fig. 31). The difference in normal maximal length (about 35 cm ♂ and 40 cm ♀) represents a considerable difference in weight (about 850 g ♂ and 1300 g ♀). Thus, because of the preponderance of older females and their larger weight, chub exceeding 1 kg are almost invariably females. Where seasonal patterns of growth have been analysed (Hellawell 1971a), it has been found that growth checks occur in May and June and are followed by a period of fast growth which decreases in September. This pattern is discernible not only in mature fish, where the check could be explained by gonad production and reproductive activity, but in immature fish also. According to Hellawell (1971b) maturity does not occur in chub of the Wye generally until the eighth year (7+), although males may mature a year earlier at some sites. In some other British rivers male maturity appears to occur much earlier (3+-5+): this difference is curious, particularly in view of the similarity in growth rates referred to earlier.

In detailed studies of diet, Hellawell (1971c) concluded that whilst older chub commonly ate fish, frogs, crayfish and even water voles, as well as a range of invertebrates and some plant material, he could find no evidence of predation on salmonids. However eels and minnows were eaten, the former being more important by volume and the latter by frequency of occurrence. In many samples all large chub had eaten similar food items, e.g. pieces of eel, crayfish or vole, and Hellawell (personal communication) suggested that such fish might possibly hunt in packs or copy each other's feeding tactics. Hellawell (1971c) also demonstrated a clear seasonal variation in feeding activity, as reflected by stomach volume: much of this variation was associated with temperature.

Turning from the chub as predator to that as prey, or potential prey, Gilbert (1929) in *The Tale of a Wye Fisherman* vividly and amusingly described the difficulties of eating chub as follows:

> We used their wretched carcasses as manure for the roses, for it is beyond the powers of most mortals to eat chub. The old description of their flesh as 'pins mixed with cotton wool' is the best, and pins and cotton wool do not appeal even to the starving. However, --- we once met an individual who could eat them in any quantity. --- We only wondered what his inside was like, and a proud fact to record is that this wonderful being was an Englishman and not a Welshman. How he must have enjoyed the War, for he could not only have digested bully-beef with ease but even the tins in which that sombre delicacy lived.

Dace

On the basis of angling returns, the dace, in many parts of the catchment, is as abundant as chub. Hellawell (1974) found that, like chub, the sex ratio changes with age towards a preponderance of females and indeed whilst some females reach an age of ten or eleven years, few males reach eight years. Unlike chub, which had one dominant age class (1959), Hellawell's work suggested that with dace years of successful recruitment occurred more frequently, a difference probably related to the earlier spawning period of this species (March in the R. Wye). Most dace of both sexes are mature in the R. Wye in their fourth year. Females when mature may have gonads weighing up to 19% of body weight.

Its growth has been determined at two sites in the upper catchment (Gough, personal communication) and is very similar to that recorded in several rivers of Southern Britain (Fig. 32) except the Thames where growth is substantially retarded throughout development (Williams 1967). Stunting of other species, such as roach and perch, was also reported by Williams (1969) in the Thames at Reading, where population densities of coarse fish (about $10/m^2$ and 500 kg/ha), particularly bleak and roach, were very high. The ultimate or maximal fork-length of dace in the Wye at the sites investigated was about 24.5 cm, equivalent to about 300 g.

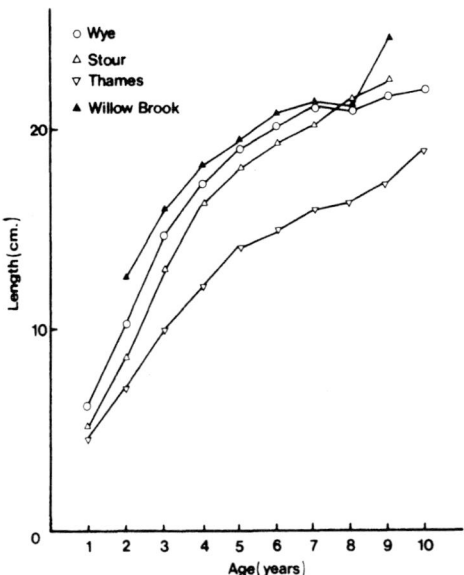

Fig. 32 Growth in length of dace in the upper Wye (W22), Willow Brook (Cragg-Hine & Jones 1969), Stour (Mann 1974) and Thames (Williams 1967).

Whilst chub have clear annual growth cycles such cycles are not distinct in dace, growth being continuous throughout the year. Feeding too, according to Hellawell (1969), is continued throughout the winter, although both the proportion of fish with food in their stomachs and the amount of food contained varies seasonally with maxima in the warmer months of late spring and summer. The diet, whilst broadly similar to that of chub, did not contain the larger items, principally fish, frogs or crayfish. Seasonal changes were evident with plant foods predominating in the summer and animal foods in the winter. Changes in diet also occurred with increasing age, the importance of aerial insects decreasing and that of caddis fly larvae increasing. Age-related changes in diet were not as dramatic as those recorded for chub, however (Hellawell 1974).

Roach

Roach are generally most abundant in the lower Wye, and on the Lugg. Hellawell (1972) was unable to obtain a sufficiently large sample on the R. Llynfi to undertake a detailed analysis of its growth: in this tributary it represented only about 1% of the larger species (Hellawell 1973). The success of spawning and early development is capricious and Hellawell found that in samples from the R. Lugg taken in 1964, containing the 1950–60 year classes, 50% were spawned in 1959.

85

Hellawell observed that growth of roach in the R. Lugg was high compared with other rivers in Southern Britain and confirmed the observation that females grow faster than males. The ultimate lengths of males and females were calculated to be about 25 and 35 cm respectively, although, with females, few seemed likely to attain that length, for it was achieved only after about 20 years and rarely did roach exceed 13 years and 30 cm. Maximum weights of the sexes reflected their lengths, being about 300 g for males and 450 g for females. These weights are considerably less than the 900 g occasionally recorded by anglers, for fish about 38 cm in length.

In the Wye roach mature when small, usually at two years, and spawning takes place during April or May in shallow water where large shoals congregate. With younger fish, males predominate (77% at 3 years), whereas with older fish there are far more females (70% at 9-12 years). Hellawell suggests that changes in the sex ratio may be caused by high mortality of females during sexual maturation and increasing male mortality with advancing age.

In analysing the diet of roach, Hellawell confirmed that feeding took place throughout the whole of the year but that there was a pronounced seasonal rhythm in stomach fullness which was associated with temperature. He pointed out that the relation between feeding rate and temperature was probably even more pronounced than that suggested by stomach volume, for, at high temperatures, digestion rates are faster, with food remaining in the stomach for shorter periods. Throughout the year, however, there was a large proportion of fish, sometimes more than half, with empty stomachs.

Plant material was the principal constituent of the diet, accounting for 60% by volume: considerable quantities of soil and detritus were also found in stomachs. Of the animals, molluscs were most important (21% by volume), *Sphaerium* and *Lymnaea* being particularly abundant. Aerial insects were rarely eaten. Hellawell's studies supported the generally accepted view that roach are essentially bottom-feeders with a preference for plants. Although there were discernible seasonal rhythms in components of diet, related to availability, these were not as pronounced as such dietary rhythms in chub and dace. There were, however, clear changes in diet with age. Young roach (0–5+) consumed more bottom-debris, macrophytes, filamentous algae and *Simulium* larvae than old roach (>10+) but did not graze diatoms so readily nor ingest trichopteran larvae. Hellawell concluded that older roach were more selective grazers, rejecting inert matter.

Pike

In 1979 pike were apparently so abundant in the Wye that the Annual Report of the Wye Division 1979 referred to a 'pike explosion' which seemed to have originated after the 1976 drought.

During 1978 and 1979 a sample (28) of pike was obtained from the lower Wye near Ross-on-Wye by fyke net and by electro-fishing, and a further sample (32) was obtained in 1977 and 1978 from the R. Elan by rod and by electro-fishing (Gough, personal communication). With such small numbers of fish, it was not possible to separate males and females in growth analysis, although it has previously been reported elsewhere that females grow faster (Mann 1976).

The growth rate of pike in the lower Wye was higher than in the R. Elan but similar to rates recorded in Windermere, and the R. Stour and R. Frome in England as well as the R. Vistula in the U.S.S.R. Faster growth rates have, however, been found in reservoirs such as Grafham Water. Gough suggests that the difference in growth rates in the R. Elan and lower Wye may be attributed to differences in the acessibility of potential prey species, cyprinids being abundant in the lower Wye and salmonids being dominant in the R. Elan. Certainly one author (Mann 1976) has commented upon the absence of salmon parr in the diet of pike in rivers where parr were relatively abundant. It seems more likely that temperature differences between the two sites account for the difference in growth rates (Kipling & Frost 1970): in the summer the R. Elan rarely rises above 14 °C, the mean being 12 °C, whereas the lower Wye achieves a mean temperature of about 17 °C. The ultimate length of pike calculated from Ford-Walford plots (Walford 1946) was about 82 cm for the lower Wye – equivalent to a weight of about 4.5 kg. This is considerably smaller than the specimens occasionally caught by anglers exceeding about 110 cm in length and 11 kg in weight.

Gough also attempted stomach analysis with the fish he caught, but few contained food (8%). Those from the lower Wye contained Twaite shad, and the one from the R. Elan contained invertebrate remains.

Grayling

Although the grayling was once included in the Salmonidae and as such held the uniquely anomalous position of being the only 'coarse-fish' member of that family, it has now been separated and forms the mono-generic family Thymallidae (Maitland 1972).

Although several workers have recorded, from scales, grayling up to 7 years, and Hutton (1923) found one on the Wye probably of 8 years, most are much younger. In the Wye Hellawell (1969) found that 86% were less than 3 years old. Growth of males and females is very similar and it appears from Fig. 33 that such growth in the Wye is slow when compared with waters elsewhere in Britain and in Scandinavia. Hellawell points out, however, that in some of these investigations the particular length measurement employed was not given and comparisons may be invalid. Seasonal variations in growth rate are different for 0+ fishes, which grow rapidly throughout the summer: other age classes after growing rapidly in the spring suffer a check in August and do not resume growth until November (Fig. 34).

According to Hellawell, maturity is attained, in the R. Lugg, towards the end of the third year (2+), spawning occurring at the beginning of the fourth year. The

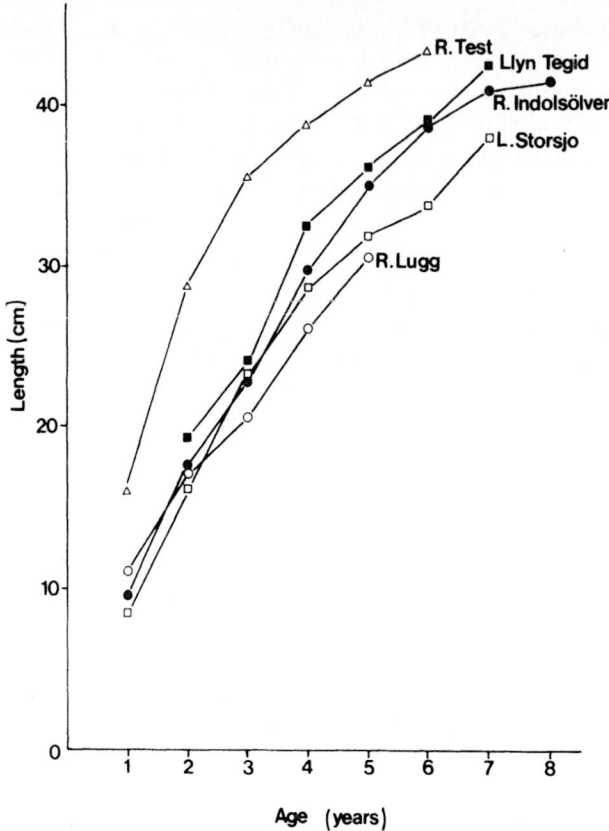

Fig. 33 Growth in length of grayling in the Rivers Lugg, Test and Indalsälven, Lake Storsjö and Llyn Tegid (from Hellawell 1969).

proportion of mature fish during the period October–December, before the period of spring spawning, increased from less than 10% in the 1+ age class to 80% for males and 100% for females in the 2+ age class.

With respect to feeding, the grayling contrasts with some cyprinids in that few fish have completely empty stomachs, indicating continuous feeding throughout the year (Hellawell 1971). If the feeding intensity is simply assessed from the bulk of food within the stomach and does not take into account seasonal changes in digestion rate, then there is a clear seasonal trend, with the lowest feeding activity during the summer months.

In contrast with the stomach contents of the chub, dace and roach from the R. Lugg which Hellawell analysed and which often contained only one constituent type, almost all grayling stomachs contained many different foods, ranging from substrate material (including pebbles, twigs and roots) and plant material (including cereal seeds and berries probably taken from the water surface) to surface foods (such as spiders and adult insects) and bottom foods.

88

Although workers elsewhere have recorded fish in the stomachs of equivalent age groups (mainly 1+ and 2+), none were found in grayling in the Wye catchment and Hellawell concluded that whilst occasional predation of other fish may occur, it is not characteristic feeding behaviour. Nevertheless the eggs of both coarse fish and salmonids were eaten occasionally. It is evident that grayling normally feed on bottom invertebrates, and in the Wye catchment this category represented about 70% by volume. Pebbles were frequently found in stomachs, as many as 45% of stomachs in April, but this seems unusual, for other workers have recorded such ingestion only rarely. Hellawell detected certain dietary changes related to age, the most obvious being the increase in importance of Trichoptera and decrease in importance of Diptera in fish in their third year.

Seasonal differences in the diet were discernible, but these were not dramatic and usually reflected changes in availability, for example aerial insects were principally restricted to the summer. There were, however, seasonal differences in the consumption of trichopteran larvae by grayling and by larger cyprinids, the former eating them in the spring, early summer and autumn and the latter eating them principally in the winter. The main winter food of the grayling in the R. Lugg was *Gammarus*.

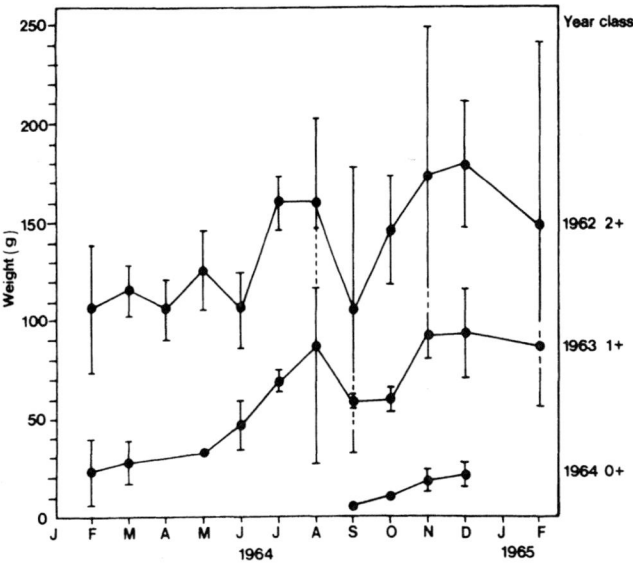

Fig. 34   Seasonal changes in growth rate of 0+ (1964), 1+ (1963) and 2+ (1962) of grayling in the Wye (from Hellawell 1969).

Bullhead

In their investigations of salmonids in the tributaries of the upper Wye, Dr. Gee and his colleagues obtained information on other species – sometimes referred to as 'forage species' – and that on bullhead, the most abundant and widely distributed, is summarised here.

Bullhead occurred at all tributary sites investigated, their average annual densities* varying between 0.25 and 2.3/m² in contrast densities on the main river were very low (<0.01/m²). Such densities, whilst higher than those recorded by Crisp et al. (1974) for the Tees river system (May densities 0.01–0.23/m², October densities 0.02–0.96/m²), are considerably lower than those from the Bere stream, a chalk river in Southern England (6.2–21.5/m² in July–August) (Mann 1971).

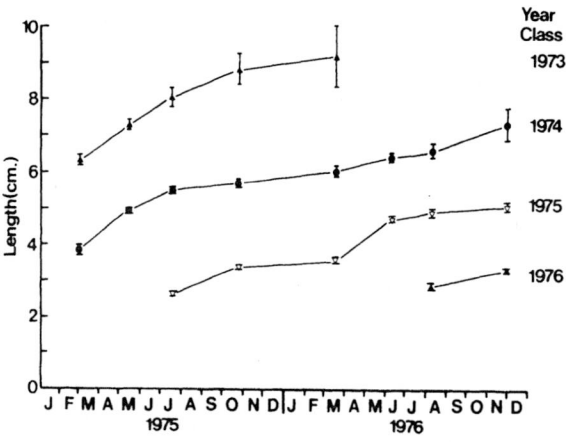

Fig. 35   Growth of bullhead in tributaries of the upper Wye.

Growth showed seasonal variation with maximum growth rates during the period March–May (Fig. 35). Appreciable differences in growth rates at various sites could be attributed to differences in bullhead density. Fig. 36a shows the relation between annual average fish density at tributary sites and the mean weight of two-year old fish: at high density sites the mean weight was only about 60% of that at low density sites. Despite this variation in size with density, average lengths are rather higher at comparable age than those studied elsewhere in the U.K. (Smyly 1957; Crisp et al. 1974) with the exception of the Bere stream where fish achieve early rapid growth but where most die after about two years.

---

* Such densities must be regarded as minimal because no depletion in numbers was obtained during successive electro-fishing runs and densities were calculated from the total number caught, normally on three runs.

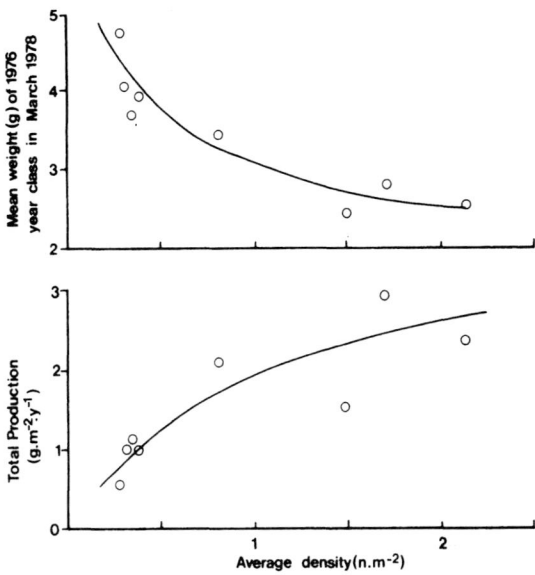

Fig. 36 Relationship between density and (a) growth and (b) production of bullhead in tributaries of the upper Wye.

The regulatory effect of density on growth reduces variation of annual production between sites (Fig. 36b), to between 0.53 and 3.1 g/m^2, when compared with variation in density. From Fig. 36b it would seem that about 3 $g/m^2/y$ is the maximum production expected for bullheads in such tributaries of the Wye. These production estimates must be regarded as minimal, however, for no account has been taken of production of eggs, and, furthermore, it is evident that 0+ fish have been grossly underestimated. In other studies such eggs and 0+ fish account for between 51 and 96% of production (Crisp et al. 1974; Mann 1971) compared with the Wye where 0+ fish contributed only between 21 and 32% of the estimated production. Despite these deficiencies in sampling younger fish, bullhead production on the Wye was apparently higher than that in the Tees catchment (0.4–1.1 $g/m^2/y$) but appreciably lower than in the Bere stream (6.2–43.1 $g/m^2/y$).

Using correlation procedures Milner (personal communication) tried to establish the degree of interspecific interaction from biomass and density data of the various fish species at tributary sites. He concluded that trout and salmon growth was unaffected by the abundance of bullheads, a conclusion tentatively reached for streams in Southern England by Mann (1971).

Brown trout

Although systematic data on the abundance, growth and movement of trout have been collected for eleven sites in four tributaries in the headwaters (Milner et al.

91

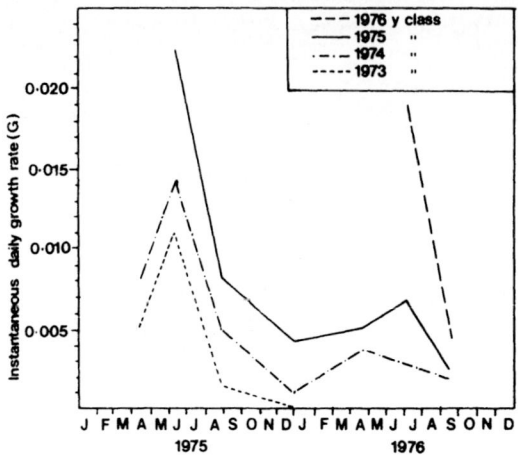

Fig. 37 Seasonal variation in instantaneous growth rate of 1+, 2+ and 3+ brown trout from the R. Duhonw near Builth Wells (from Milner et al. 1978).

1978, 1979), the general features of trout behaviour in the catchment cannot be described. In the main river and its larger tributaries native populations are extensively supplemented by stocking: these introduced fish are frequently larger than most native ones (<0.5 kg).

Maximum densities in the sites varied between 0.06 and 0.28/ m². Only four-year classes were generally found, 4+ and 5+ trout being uncommon. Age structure of the site populations clearly changed seasonally with the emergence of fry from the gravel in May and June. Immigration and emigration, established from dye-marking, was also a significant feature at most sites with about 30% of unmarked fish being found within site samples every three months – some of these may have travelled only short distances, of course. Some seasonality in movement was

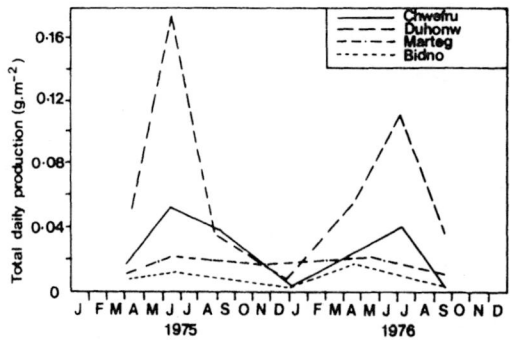

Fig. 38 Seasonal variation in production at sites on the Rivers Duhonw, Chwefru, Marteg and Bidno (from Milner et al. 1978).

92

established, 1+ and 2+ trout being more mobile during the spring and early summer. These immigrant fish, as has been shown by other investigators, emigrate more readily than residents. Milner et al. (1979) calculated the time for half the marked trout at sites to be exchanged: these times varied between 140 and 430 days.

Growth rates are not only markedly seasonal and age dependent, they also vary considerably between years, and Fig. 37 shows the features of growth, expressed as instantaneous daily growth rate, in one of the tributaries, the Duhonw. Production, being a product of biomass and growth rate, varies seasonally (Fig. 38). At most sites Milner et al. (1978) found that 0+ fish contributed less than 10% of total production and that immigration of all age classes did not represent a high proportion (generally 30%) of site production. Annual trout production, which was related to biomass, varied between 2.9 and 19.7 g/m², the latter value being high for upland streams in Britain. When salmon production at these sites is taken into account, total salmonid production ranged from 7.7 to 26.8 g/m²/y – generally well above the value of 12 g/m²/y regarded by Le Cren (1969), from a literature survey, as a normal maximum for streams.

Trout production was not determined in the richer lowland tributaries, but, within this limited range of upland sites studied, there was a correlation between annual trout production (g/m²) and mean calcium concentration (mg/l) (Fig. 39), the regression equation being

$$\log_{10}P = 0.562 + 0.033\,[Ca^{++}]\;(r = 0.85).$$

Production seems to conform to the same relationship when data from other U.K. river systems are included (Milner, personal communication). In the Wye tributaries most of the variation in production was related to fish density or holding capacity. The physical characteristics of the streams determining holding capacity

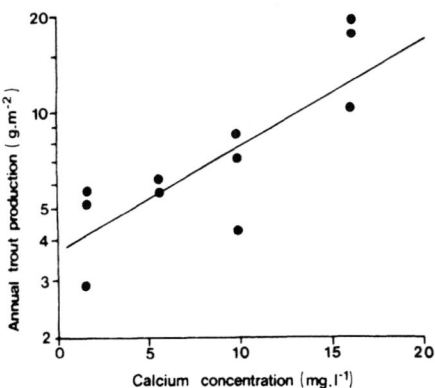

Fig. 39 Relationship between trout production and average calcium concentration at sites in the upper Wye catchment.

include gradient, substrate stability, structure of riffles and pools and bankside cover. These are in turn determined largely by the geology of the catchment, which concomitantly influences water quality. The apparent relationship between calcium and fish production is therefore probably mediated through geological characteristics influencing the physical features of the riparian habitat.

Salmon

The salmon is of particular interest in the Wye, not only because it represents the most important fishery resource of the river (Randerson et al. 1977), but also because its history of exploitation has been particularly well recorded and because certain changes in the age and size distribution of the stock seem explicable in terms of changes in that exploitation.

Eggs and juveniles. Spawning gravels are extensively distributed throughout the Wye catchment, but not all sites where fish regularly spawn are equally suitable. Fertilised eggs were planted in plastic boxes within gravel (Milner, personal communication) at ten sites in the upper Wye and its tributaries, and, with the exception of two sites on the R. Claerwen and in the headwaters of the R. Wye, these sites were chosen where salmon regularly spawn. Although survival was greater than 80% at some sites, at others it was appreciably lower (67–47%). Low survival was generally associated with low concentrations of dissolved oxygen ($<$6 mg/l) in intra-gravel water, a characteristic related to low permeability of the spawning gravel. Where there were low survival rates in egg boxes, the specific locations where the boxes were buried may not have been those normally chosen by salmon.

The average median hatch time for all sites was 125 days post-fertilisation at average incubation temperatures of 2.5 to 3.5 °C, and this development period seems to conform with the relationship between hatching period and temperature proposed by Crisp (1981). From the normal times of spawning activity in the river, it seems that hatching occurs over an extensive period, between February and May, with peak hatching in late March. As with developmental period, the growth rate of embryos (expressed as increased in length) was, from the five sites where maximum and minimum temperature measurements were recorded weekly, closely related to mean temperature: 86% of the variance in growth rates was accounted for by temperature variation between sites (Fig. 40).

The production and population dynamics of juvenile salmon have been studied at many sites in the upper Wye catchment by Gee et al. (1978a, b). In all but the coldest headwaters parr generally spend only two years before becoming smolts. Gee (personal communication) calculated that about 12% of smolts were 3+ in the R. Bidno, the highest tributary he investigated, and appreciable numbers of 1+

94

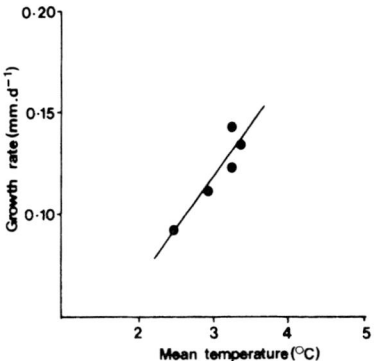

Fig. 40 Relationship between growth rate of embryos of salmon and mean water temperature at sites in the upper Wye catchment.

smolts have been reported in the lower reaches of the river. Scale analysis suggests that between the periods 1908–39 and 1962–77 the proportion of 1+ smolts increased and of 3+ smolts decreased. This shift is not entirely explicable in terms of climatic changes, however.

The R. Wye, with its considerable length and large and diverse catchment, provides a wide range of environmental conditions which leads to variation in smolt age and facilitates the spatial segregation of spawners. Both factors tend to maintain genetic differentiation of the stocks and provide the succession of runs for which the river is renowned.

During the period from June 1st, soon after the emergence of fry from the gravel, until smolt migration about two years later, the instantaneous loss rate (mortality and net emigration) for a given cohort, or age-class, at a particular site appeared to remain constant. Furthermore this loss rate was dependent on the initial fry density (Fig. 41a). In consequence of the 'over-compensatory' nature of this relationship, the maximum number of pre-smolts per unit area, about $0.04/m^2$, was derived from an intermediate fry density of about $0.75/m^2$ – lower and higher densities of fry producing lower yields of pre-smolts (Fig. 41b). At this optimal density, survival was about 4%. The mechanism whereby initially high densities of fry maintain a high loss rate, either through emigration or mortality, even when these densities are reduced below those of populations with lower initial densities and lower loss rates, is by no means clear. Furthermore, the relationship between initial fry density and subsequent mortality seems very different in the case of brown trout, for Le Cren (1973) found that over a range of initially high alevin densities the number of survivors was constant, perfect density-dependent regulation. Another interesting feature of juvenile salmon survivorship as reported by Gee et al. (1978a) is that mortality seems uninfluenced by the presence and abundance of co-existing fish species, suggesting perhaps fairly rigorous partitioning of the habitat.

95

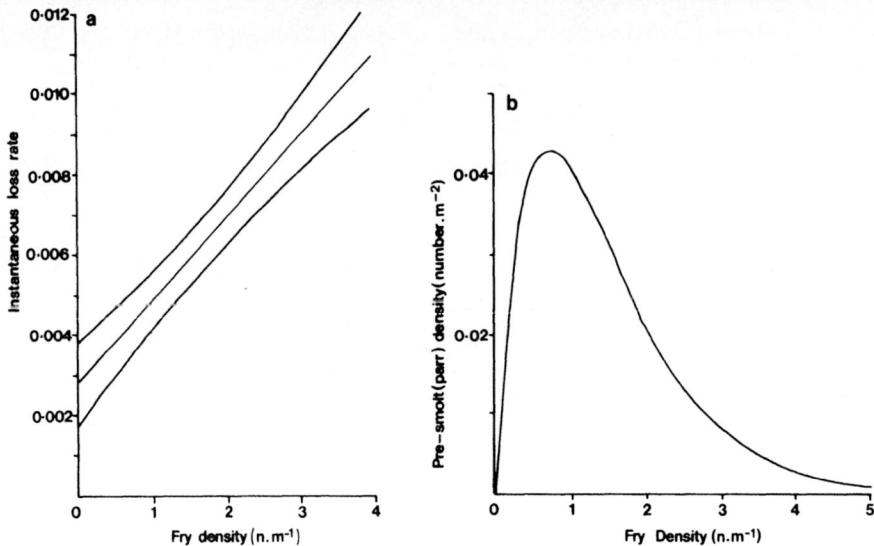

Fig. 41 (a) Relationship between instantaneous loss rate and estimated fry density on June 1st in tributary sites: 95% confidence limits for the regression line are shown. (b) Relationship between estimated pre-smolt densities and estimated fry densities on June 1st (from Gee et al. 1978a).

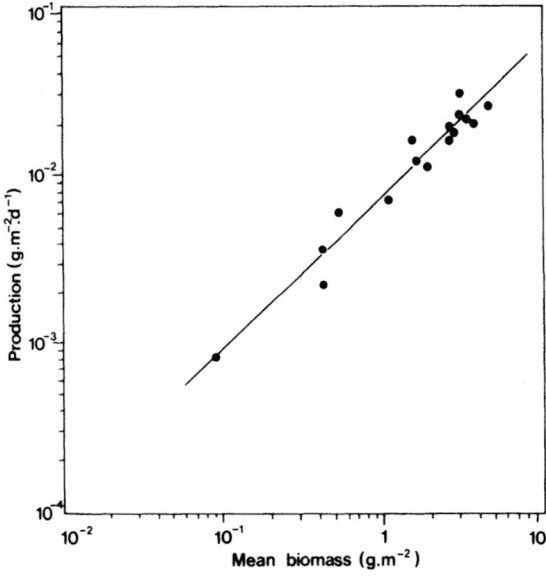

Fig. 42 Relationship between production and biomass for salmon at sites in the upper Wye catchment (from Gee et al. 1978b).

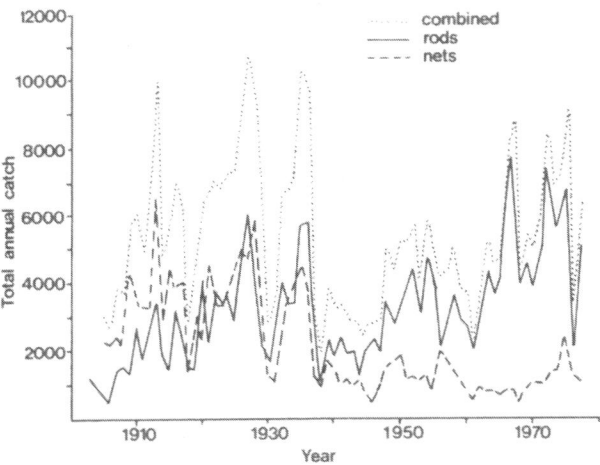

Fig. 43 Annual salmon catches from rods, nets and combined methods (from Gee & Milner 1980).

The production of juvenile salmon may also be expressed in terms of biomass and, at the sites which Gee et al. (1978b) investigated, such production was within the range 0.3 to 11.0 g/m²/y, a range similar to that recorded by Egglishaw (1970) for a Scottish stream which was studied intensively. As for brown trout, Gee et al. (1978b) concluded that production at sites was principally determined by biomass (Fig. 42), the growth rate varying little between sites; in consequence, those factors which limit the biomass limit production. In the tributaries of the upper Wye it was concluded that the number of spawners reaching sites, related partly to accessibility, was an important feature in determining salmon production.

Adults. The numbers of adult salmon ascending the Wye must be inferred from those which are caught by commercial and sport fishermen, for no trap nor counter has been installed near the estuary.

In the nineteenth century, widespread netting of migrants in the middle and lower reaches of the river, intensive spearing and gaffing by land-owners and poachers on the spawning beds, and the capture and removal of salmon parr (samlets) for eating, decimated the salmon stocks. Although the widespread slaughter continued until the end of the ninteenth century, a nucleus of fishery owners established a Wye Preservation Society one year after the Salmon Fishery Act of 1861. From this Society, which had no legal status, developed the Wye Fisheries Association and Wye Conservators. The latter organization, whilst effective in catching poachers, ignored the equally damaging freshwater and estuarial netting. The greed of the netsman led to their undoing for in the last decades of the century their catches were severely reduced, and netting rights were sold for modest sums to the Wye Fisheries Assocation. All netting in the river and estuary became

97

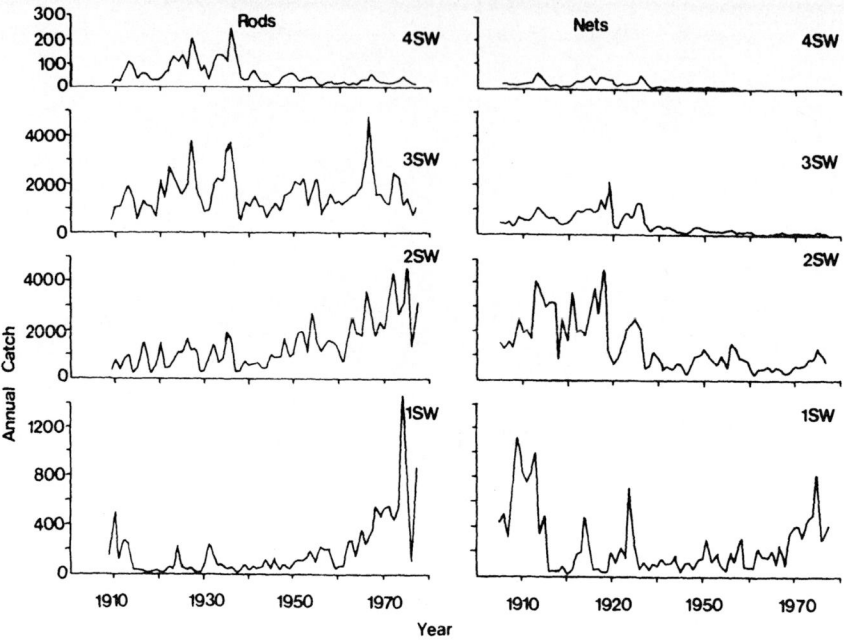

Fig. 44 Annual catches of rods and nets of salmon in sea-age groups by weight (from Gee & Milner 1980).

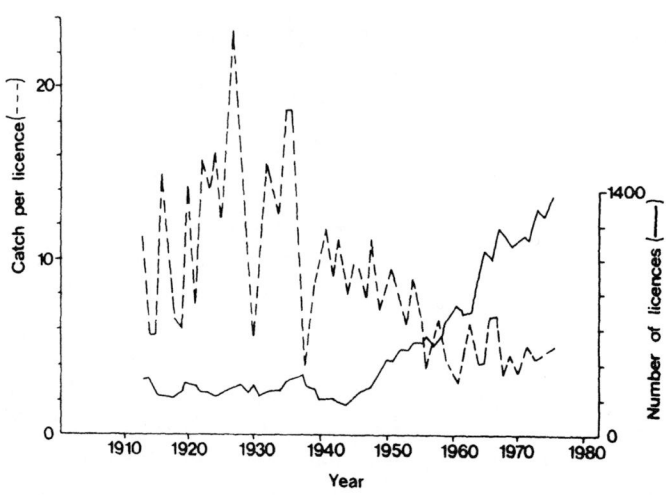

Fig. 45 Changes in rod fishing effort (as expressed in whole season licences) and catch per unit effort (from Gee and Milner 1980).

controlled in 1901 when the last netting rights were leased from the Crown which had purchased them from the Duke of Beaufort.

After a three-year ban on estuarial fishing and the severe reduction in fishing in the river, catches began to recover, and total annual rod and net catches of about 6000, similar to those of today, were recorded within a decade. Since the early years of the century, one of the major changes has been the increased proportion of the catch taken by rods - now about 80% of the total catch. The only net fishery, in the estuary, is operated on a non-profit basis by the statutory fishery agency (now the Welsh Water Authority).

With this history of the salmon fisheries and their control by very few owners, about one hundred, the record of catches has remained exceptionally good. These catches, from 1905 to 1977, have recently been analysed by Gee and Milner (1980) in relation to the changing pattern of exploitation.

From 1910 to 1940 total catches by both rod and net fisheries were similar (Fig. 43). Since then the annual net catch has decreased and remained at an average of about 1000 whilst the annual rod catch has increased, generally to between 4000 and 7000. During the period 1909-77, the annual catches of both grilse and 3SW*, but not 2SW fish, caught in nets and by rods were positively correlated. This suggests that there is no immediate competitive effect of the net on the rod fishery and that, in the long-term, catches of salmon by both methods may be affected by common factors such as stock abundance and environmental conditions.

In both rod and net fisheries the average weight of fish has decreased, and when the catches are examined in terms of age or weight categories, the reason for the overall weight change becomes evident (Fig. 44). Particularly since the period 1940-50 the proportions of fish which have remained at sea for only one or two winters (1 and 2SW fish) have increased in both the rod and net fisheries. In the rod fishery, the proportion of 4SW fish has decreased from 4 to 0.5% and that of 3SW fish, the dominant age group before 1945, from 60-65 to less than 30%: 2SW fish have increased from 30 to 60% and 1SW fish (grilse) from less than 5 to a peak of 25% in 1974. In the net fishery large fish were never dominant because fishing was not undertaken when many of these large fish moved into the river during the spring.

Gee and Milner (1980) have suggested that the change in age-composition of the catch results from the increased exploitation, reflected in the seven-fold increase in rod licences since 1945 (Fig. 45). Whilst the catch per licence has decreased, there has been an overall increase in the rod catch of salmon (Fig. 44) principally of the younger age classes. It has been estimated that at current exploitation rates whilst less than 20% of grilse are caught by rod more than 98% of 3 and 4SW fish are similarly caught. Gee & Milner (1980) suggest that at such high exploitation rates of large fish   caused principally by their early entry into the river and their

* Salmon which have spent two, three and four winters at sea are referred to as 2SW, 3SW and 4SW fish respectively.

99

availability for most of the fishing season – the spawning stock is predominantly composed of grilse and 2SW salmon. There is recent evidence of a genetic basis to the duration of sea absence (Gardner 1976), and therefore it seems likely that the reduced proportion of large fish available for spawning influences the age and size distribution of the progeny.

Other possible causes of the change in age composition of the catch have been suggested. That most frequently proposed, the Greenland salmon fishery, seems unlikely as its considerable expansion in the early 1960s postdates catch changes in the R. Wye.

A computer simulation model (Gee & Edwards, in press; Gee & Radford, in prep.) has been developed for salmon stocks of the R. Wye which can explore the impact of changes in mortality rates, including fishery exploitation, on stock abundance and composition. The model calculates the number of survivors from a time series of six cohorts throughout their life cycle. As fish spend variable times at sea this feature must be incorporated in the model. The size of each cohort is determined by the number, size distribution and sex ratio of the parent stock of spawners and losses during embryo development are taken from Symons (1979). Juvenile mortality is assumed to be density dependent, as described by Gee et al. (1978), and survival rates at sea are selected from tagging returns for wild British smolts, these being 5.3, 2.7 and 1.6% for 1, 2 and 3SW salmon respectively.

Fig. 46 shows two sets of assumptions concerning the stock of salmon and its exploitation. In Simulation 1 it has been assumed that all smolts have an equal chance of becoming 1, 2 and 3SW fish, differences in numbers of returning adults resulting solely from continued sea mortality of older fish and that no exploitation of stocks occurs within the estuary and river. Within 7 years of stocking with 2 million eggs ($\equiv 0.13/m^2$), the total adult population had stabilised around 18 500, comprised of 10 300 1SW, 5200 2SW and 3000 3SW salmon. By introducing a random element, from year to year, in mortality at sea, an oscillation with a periodicity of about a decade, equivalent to twice the life cycle, was detected. Such a periodicity in abundance was postulated by Ricker (1954). In Simulation 2 two modifications were applied. It was assumed, firstly, that there was a partial genetic basis to sea-absence (Gee & Radford, in prep.) and, secondly, that exploitation within the river approximated to current rates (5, 50 and 90% for 1, 2 and 3SW fish) which were different for each age-group and dependent on average residence time in the river within the fishing season. After falling initially with the onset of exploitation, recovery occurred, and the run increased to 25 000 (about 6500 above that without exploitation). However, the age-composition of the spawners was very different from that without exploitation, with an increase in the proportion of grilse to 2SW and 3SW salmon from about 55 to 85%. This response to exploitation in the simulation model has features very similar to those observed in the fisheries of the R. Wye during the last few decades. Whilst this similarity suggests that the cause for changes in age composition postulated by Gee and Milner (1980) is plausible, similar patterns of differential exploitation with age during the marine

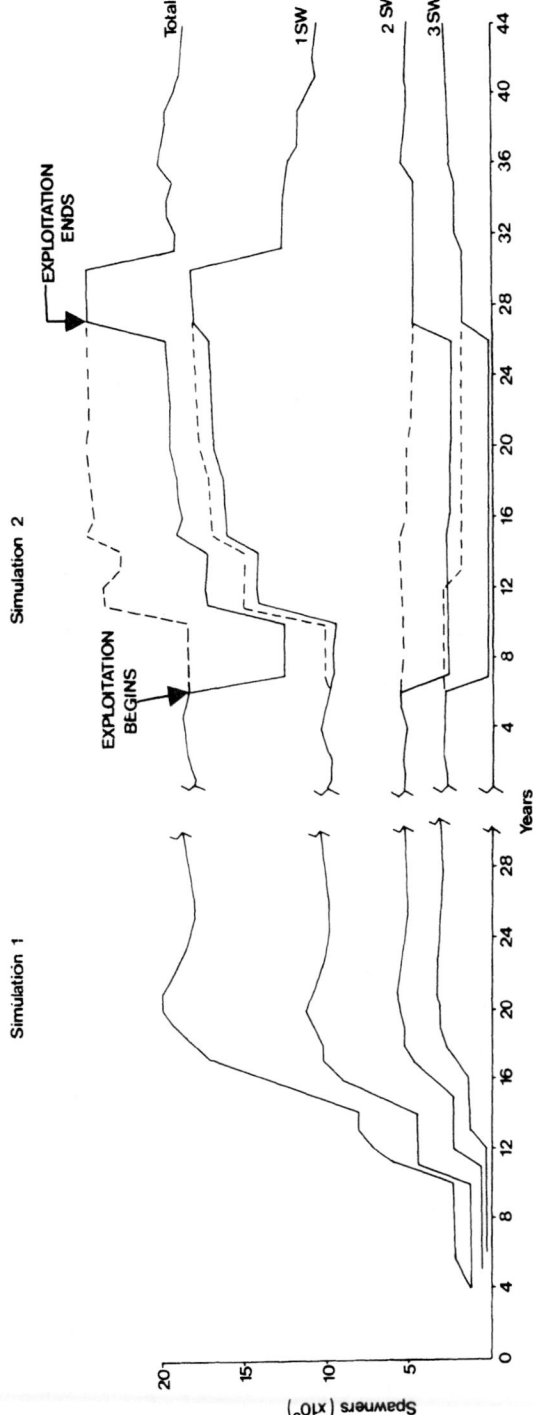

Fig. 46 Numbers of spawning salmon in the R. Wye according to the simulation model (for details see text). In Simulation 1 there is no exploitation by rods and nets and in Simulation 2 exploitation (at different levels for each age class) occurs. The hatched lines represent the total numbers of salmon returning to the river before exploitation.

101

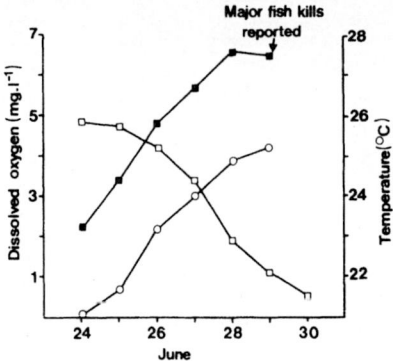

Fig. 47 Changes in water temperature and dissolved oxygen concentration near Ross during the period 24–30 June 1976 (from Brooker et al. 1977). (–□– min. dissolved oxygen; –■– max. temp; –O– min. temp).

phase of their life-history would have a similar effect. The response of the model to the cessation of exploitation was rapid, with a return to the abundance and age composition of the stock characteristic of 'pre-exploitation' within four years (Fig. 46). Whether the temporary control of the exploitation of large fish is regarded as worthwhile in seeking to increase their numbers is as much a matter of political, as of scientific, judgement.

In addition to the exploitation of salmon within the R. Wye, there have been substantial freshwater mortalities, particularly during the late 1960s, resulting from ulcerative dermal necrosis. The quantitative effect of this disease on stocks has not been assessed, although it has been estimated that in 1968 and 1969 respectively about 2500 and 1100 salmon were killed, albeit many of them kelts on the spawning beds.

In 1976, a year of substantial summer drought, when flows in the lower river were only between 20 and 30% of the long-term average during June and July, many salmon died in a 30 km reach between Ross and Monmouth where growths of water crowfoot (*Ranunculus fluitans*) were prolific (Brooker et al. 1977). The mortalities were coincident with high river temperatures and with the decay of the plant stand, the latter condition inducing low oxygen concentrations (Fig. 47).

6. Birds and mammals

Birds

Like other areas of Britain, observations of birds in the counties through which the Wye flows are regularly recorded and published, although the shortage of field ornithologists in Wales is reflected in the paucity of records in this area, and no systematic survey of the Wye had been undertaken until the spring and summer of 1977 when the Royal Society for the Protection of Birds was commissioned by the Nature Conservancy Council to undertake a detailed survey of breeding birds. Apart from the Waterways Bird Survey (Williamson 1975) which covered 250 km of rivers in 1974 and 374 km in 1975, this was the first extensive birds survey of a major river system in Britain not confined to one or a limited group of species. The main objectives of the survey were to determine the distribution of birds using, and breeding along, the river and to relate this distribution to the spatial differences in the character of the river and the land through which it flows: prediction of the effects of river management, particularly in relation to discharges arising from an enlargement of Craig Goch Reservoir, was also required. This chapter is based largely on the RSPB survey, supervised by C. J. Cadbury and R. Lovegrove and undertaken by A. Merritt, M. Moss and C. Walker, which involved identifying and counting breeding pairs of birds on the Wye, from the source to Redbrook (W61), and on adjacent sections of tributaries and nearby pools, between the end of March and mid-July 1977.

The river seemed to divide naturally, from the viewpoint of bird habitats, into seven sections (Fig. 48).

Section I (W1–W13):
(fast-flowing, contains boulders, riffles and shingle banks, adjoining land is mainly used for rough grazing).

Section II (W14–W24):
(fast-flowing, with banks which are extensively wooded, although there is some pasture and rough grazing).

Section III (W25–W26):
(flowing more slowly with few riffles and rapids but some shingle and earth banks, land used mainly for dairy farming).

103

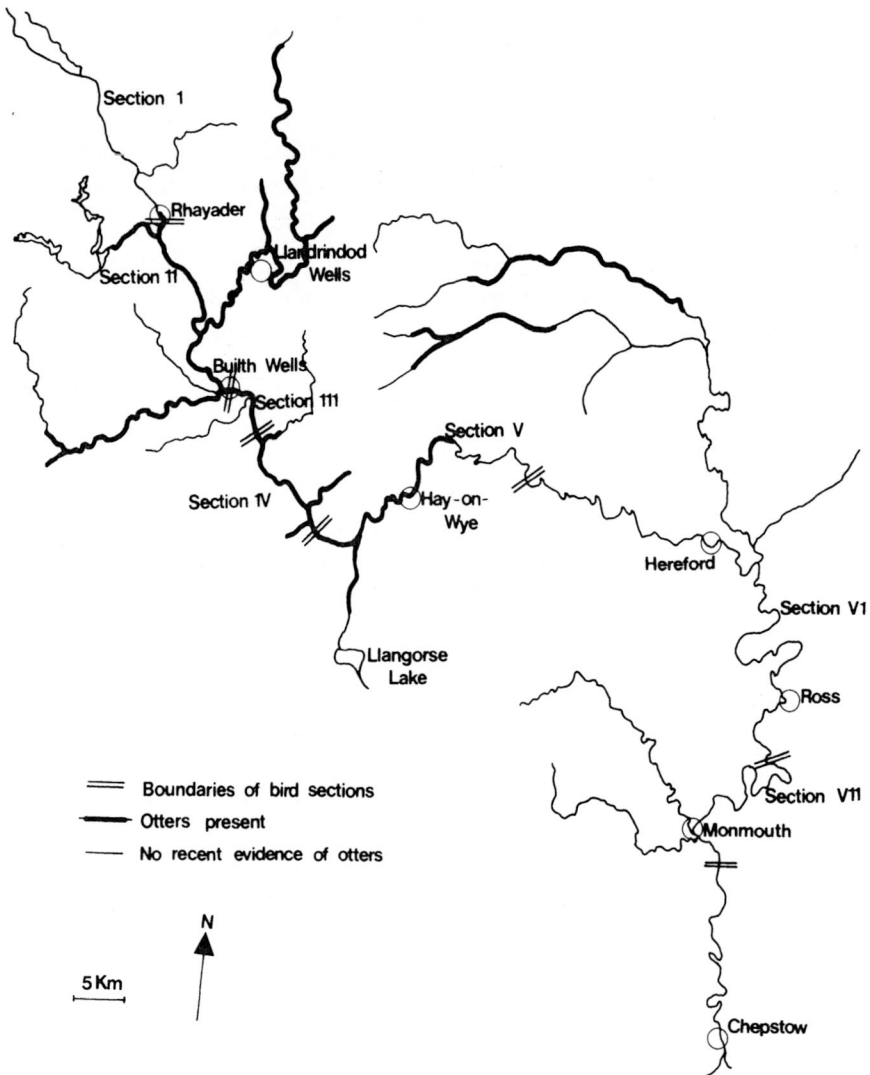

Fig. 48 Map of the Wye showing the distribution of otters and boundaries of the seven sections distinguished in the bird survey (see text for details).

104

Section IV (W27-W30):
(rapid, with exposed boulders and shingle, banks generally wooded).

Section V (W31-W43):
(many riffles, earth- and shingle-banks, adjoining land used primarily for dairy and arable farming).

Section VI (W44-W54):
(slow-flowing, with many earth-banks and some cliffs at Bredwardine and Holme Lacey, land used mainly for dairy farming).

Section VII (W55-W61):
(slow-flowing except for short stretches near Symond's Yat, apart from the well-wooded gorge in the Forest of Dean, the adjoining land used for dairy and arable farming).

During the survey 129 species of birds were recorded (Appendix 5), 71 breeding on the banks of the river. Of these recorded species, 58 were censused (Table 20) and could be classified as follows: A) closely associated with the river when breeding, B) used the river but did not nest there, C) waders nesting on adjacent water meadows and marshes and D) associated with marginal scrub and woodland. Consideration will now be given to these groups.

River breeders. Some species, such as mallard, moorhen, kingfisher, pied wagtail and grey wagtail, are fairly evenly distributed throughout most of the river system. In 1977 there were about 300 pairs of mallard breeding on the Wye, and the density of males exceeded 10 per 10 km except in the fast-flowing upland reach above Rhayader (Section I). Around Hereford there was a large feral population and some hybrids with domestic birds: these were excluded from the census. The moorhen, although rarely observed and not breeding in the upper reaches (Section I), was common along most of the Wye, densities being similar to those recorded in the Waterways Bird Survey (Williamson 1975). The kingfisher was widely distributed in the catchment except in Section I, which has an altitude range above that normally occupied by this species (<180 m) and few earth-banks suitable for nesting. Kingfishers were most frequently found near fast-flowing reaches with exposed boulders; they also favoured tributary junctions within their territories and often for their actual nest sites. The pied wagtail was one of the commonest birds on the Wye and, although its nests are generally some distance from the river, often around buildings, shingle-banks on the river were extensively used as feeding areas. The grey wagtail is generally most abundant near fast-flowing rivers, achieving densities up to about 15 pairs per 10 km on the R. Ogwen in North Wales (Williamson 1975). On the Wye its densities were generally lower and were predictably highest in reaches with high water velocities, predominantly in Sections I, II and IV.

Table 20 Density of territories (per 10 km) of species closely associated with the river corridor.

	Sections						
	I	II	III	IV	V	VI	VII
Species closely associated with the river							
Great-crested grebe					1.0		
Mallard	3.1	12.8	18.7	12.0	8.3	15.0	6.7
Tufted duck					P		
Red-breasted merganser		P					
Goosander	P	B	P	P	P	P	
Mute swan					P	0.7	0.3
Moorhen	P	2.0	5.3	6.4	8.5	13.0	5.0
Coot	P	P	P	2.7	1.4	0.7	
Common sandpiper	8.0	8.8	5.3	16.0	7.8	0.7	0.3
Kingfisher	0.3	3.2	4.0	2.4	2.8	2.6	2.0
Sand-martin	24.6	7.2	100.0	20.0	247.0	151.0	16.7
Dipper	4.9	7.9	1.3	6.4	0.5		
Sedge warbler		1.2			6.4	5.6	2.3
Pied wagtail	6.8	6.4	10.7	9.6	10.1	16.7	12.7
Grey wagtail	5.9	6.8	4.0	11.2	2.8	0.9	2.7
Yellow wagtail					12.0	18.6	2.3
Reed bunting	0.6	P			10.4	15.3	5.3
Species which used the river but did not nest on it.							
Little grebe				P	P	P	P
Cormorant		(2.0)			(8.5)	(4.7)	(0.3)
Heron	(2.8)	(15.9)	(18.7)	(20.8)	(20.3)	(6.3)	(0.7)
Teal					P		
Pochard					P	P	P
Goldeneye					P	P	P
Canada goose					P	P	
Green sandpiper					P		
Greenshank					P		
Great black-backed gull					(0.5)	(0.1)	
Lesser black-backed gull		(7.2)	(10.7)	(8.0)	(9.4)	(13.6)	(6.7)
Herring gull					(1.2)	(0.4)	(2.3)
Common gull						(0.3)	(0.3)
Black-headed gull	(2.8)	(2.8)	(16.0)	(10.4)	(19.1)	(7.0)	(1.3)
Waders which bred (or could breed) on land adjacent to the river							
Lapwing	1.5	P			3.8	5.4	1.3
Little ringed plover						B	
Golden plover	B				P		
Snipe					P		
Curlew	4.6	1.6	6.7	0.8	5.7	1.6	
Redshank					P		

Table 20 (continued).

Species associated with scrub and woodland on the banks							
Buzzard	(11.1)	(13.6)	(16.0)	(9.6)	(3.3)	(1.0)	(1.7)
Sparrowhawk	(2.2)	(2.0)		(4.0)	(1.2)	(0.9)	(1.3)
Kestrel	(1.9)	(4.0)		(1.6)	(0.2)	(3.9)	(1.0)
Woodcock	B	B					
Little owl	(0.3)	(1.2)			(1.9)	(1.9)	
Green woodpecker	(1.9)	(2.4)	(1.3)	(1.6)	(1.9)	(5.4)	(4.0)
Great spotted woodpecker	(3.1)	(7.2)	(1.3)	(11.2)	(6.1)	(8.6)	(9.0)
Lesser spotted woodpecker		(0.8)			(0.2)	(1.4)	(0.3)
Marsh tit	(1.2)	(2.8)	(1.3)	(2.4)	(0.2)	(0.4)	(1.0)
Willow tit	(0.3)	(0.4)		(0.8)		P	P
Nuthatch	(2.5)	(8.8)	(1.3)	(12.0)	(1.2)	(0.7)	(1.3)
Tree-creeper	(5.9)	(15.6)	(10.7)	(15.2)	(3.3)	(1.9)	(3.0)
Whinchat	(2.5)	(0.8)			(0.2)	(0.1)	
Redstart	(5.9)	(6.0)	P	(1.6)	(0.2)	(0.1)	
Blackcap	(1.9)	(9.2)	(9.3)	(6.4)	(9.7)	(10.4)	(14.7)
Garden warbler	P	(2.4)	(4.0)	(4.0)	(1.7)	(0.7)	(1.7)
Whitethroat	P	(3.2)	(4.0)	(0.8)	(4.7)	(6.0)	(6.3)
Lesser whitethroat					(1.2)	(0.3)	(1.3)
Spotted fly-catcher	(2.2)	(9.6)	(1.3)	(7.2)	(2.8)	(3.7)	(3.0)
Pied fly-catcher	(3.7)	(6.4)		(16.0)		P	

P denotes presence only and B evidence of breeding. Figures in brackets denote pair densities of species which do not necessarily nest on the river bank

Only two species were markedly associated with the upper reaches of the Wye, namely the common sandpiper and dipper. The sandpiper was restricted to areas where there are riffles and exposed boulders and shingle-banks on which the bird feeds. The dipper was even more restricted in its distribution (Sections I–IV). Several nests were found under bridges and were sometimes so close to the water that flooding must present a serious hazard.

Several species were found principally in the lower reaches of the Wye (coot, mute swan, sedge warbler, reed-bunting, yellow wagtail, sand-martin). The coot is generally regarded as a bird of still or slow-moving weedy waters and in the Wye its territories were most frequently on pools adjacent to the river. The mute swan nested only infrequently on the river and of the six pairs which did so in 1977 only one pair successfully raised young. Most of the broods which were found on the river had moved from nearby pools or tributaries. The sedge warbler, a bird of emergent vegetation and bank-side scrub, was confined to downstream reaches where this habitat was abundant (Sections IV to VII). The reed-bunting, similarly distributed on the Wye, was found principally in reaches with hedgerows and scrub dominated by willow, alder and hawthorn. Towards the end of June several male reed-buntings were recorded in the upstream reaches of Section I, and it is likely that these were unpaired males attempting to colonise marginal scrub in the area.

The yellow wagtail, a bird which favours damp grasslands for breeding, has a limited world distribution in western Europe, centred on Britain and Ireland. The total population in Britain and Ireland is only around 25 000 pairs (Sharrock 1976) and is considered to be declining in Southern England as a result of land drainage. In the Wye it is near the western limit of its breeding range and this may account for the absence of territories in Section III which otherwise seems suitable, with open pasture and dairy farming: dung-flies, which are likely to be abundant in such areas, represent an important item of diet for yellow wagtails. In the Wye about 40% of pairs of yellow wagtails were associated with group, rather than individual, territories, and these were in areas of high abundance downstream of Section V. The sand-martin which requires both earth-banks for nesting and open country for feeding, was relatively abundant in Sections III, V and VI, but large nesting colonies, consisting of above 50 nests, were only found in Sections V and VI. The average nest height was about 2.2 m above water level, holes being excavated near the top of banks, and nests seem in little danger from normal flooding.

In addition to these species with extensive distributions, there were some (great-crested grebe, tufted duck, red-breasted merganser, goosander) which occurred only in low numbers and therefore were only found in limited areas. Great-crested grebes, normally found on shallow ponds and lakes rather than rivers, were recorded on slow-moving stretches in Section V in territories containing willows and generally an abundance of *Phalaris*. Of the four pairs, three successfully reared young in 1977. Four pairs of tufted duck were also seen in Section V, although, of these, two pairs were on an ox bow lake near the river. This is a late breeding species, and the absence of broods during the period of observation in 1977 may not have indicated breeding failure. There were almost 30 sightings of the red-breasted merganser near Newbridge-on-Wye (Section II), and, although breeding was not confirmed, the dense ground cover seemed most suitable for nesting and, being another late breeding species, broods would not have been seen during the survey period. Wales has been only recently colonised by this species (1953), but its numbers have been steadily increasing. Two breeding pairs of goosanders were also located in Section II, a heavily wooded reach favoured by this species which often nests in hollow trees, but birds were observed on most sections of the river and some of these seem suitable for breeding. This species, confirmed as a breeder in Wales as late as 1970, now has several breeding pairs on upland reservoirs and the higher reaches of rivers in many parts of North and Mid-Wales.

River users. Of those species which use the river, some, such as the little grebe, pochard, Canada goose and black-headed gull, breed on adjacent pools but not on its banks: the heron has nesting colonies containing up to about 200 nests within 25 km of the river. The gulls and cormorants, with a predominantly coastal distribution, tend to occur most frequently in lowland reaches. Greenshank, green sandpiper and teal are regular visitors to the river, the goldeneye a more occasional one.

Waders. Most of the recorded species of waders which might normally be expected to nest on land near the river did so, although numbers were not high; in the case of the curlew and redshank, there was no evidence of successful breeding, and this is probably caused by the extensive drainage of grassland. Of particular note was the successful breeding of a pair of little ringed plover on a gravel pit near Hereford – the first breeding record for the Wye valley of a species with a breeding population in Britain of only about 500 pairs.

Species of marginal scrub and woodland. As might be expected, the widest variety of bird species recorded during the survey used the woodland corridor along the banks of the Wye. Although most were widely distributed – particularly in heavily wooded reaches – a few were confined to, or more numerous in, the upper reaches (buzzard, woodcock, whinchat, redstart, pied fly-catcher), others to the lower reaches (lesser spotted woodpecker, lesser whitethroat) and several, such as the blackcap, garden warbler and whitethroat, although widely distributed, were least abundant in the open uplands above Rhayader (Section I).

Of these species, the willow tit and lesser whitethroat were locally scarce, and the restricted distribution of the latter on the lower reaches may well be associated more with its being at the western extremity of its range than any concentration of its preferred habitat, the hedgerow, in these reaches.

Effects of river management

During the survey, observations were made on adjacent reaches of river with different degrees of river management, and, although results need cautious interpretation, differences in bird populations were so large that some explanation seems necessary. In one sequence of reaches near Hay-on-Wye (Section V), unmanaged sections had not been dredged and were meandering. There were shingle islands and riffles and the river banks were shelving, supporting extensive stands of tall grasses and willow scrub, or were high earth-banks. Adjacent land was mainly rough pasture. In contrast, the managed section had been dredged – with the removal of shingle banks – and the river was sluggish and uniform in flow pattern. The banks were graded and lined with trees, but no scrub and the adjacent land, well tended, was used for grazing and arable crops. The managed reach had only 11 breeding species (43 territories/ 10 km) compared with 25 breeding species (146 territories/ 10 km) in adjacent unmanaged areas. The most striking differences in distribution and abundance (Table 21) were in mallard, mute swan, moorhen and coot, which were generally absent from managed areas. Others particularly affected were common sandpiper, which feeds on shingle banks and islands; sand-martin and kingfisher, which nest in earth banks; sedge-warbler, which requires tall, bank-side vegetation; and the lapwing, redshank and curlew, which frequent poorly drained pastures. Populations of warblers and other passerines

Table 21 Density of territories of species in managed and unmanaged reaches of the R. Wye near Hay (Section V).

Type (see p. 105)		Territories (number per 10 km)	
		Unmanaged	Managed
1	Great-crested grebe	1.4	–
	Mallard	4.3	–
	Mute swan	2.9	–
	Moorhen	8.6	2.9
	Coot	4.2	–
	Common sandpiper	15.7	–
	Kingfisher	2	–
	Sand-martin	50	10
	Dipper	1	–
	Sedge warbler	11.4	–
	Pied wagtail	11.4	5.7
	Grey wagtail	4.3	2.5
	Yellow wagtail	13.0	5.7
	Reed bunting	14.3	8.6
2 & 3	Heron	(8.6)	(5.7)
	Curlew	7.1	2.9
	Lapwing	10.0	–
	Redshank	1.4	–
4	Nuthatch	2.9	2.9
	Tree-creeper	5.7	–
	Whinchat	1.4	–
	Blackcap	11.4	5.7
	Garden warbler	2.9	–
	Whitethroat	4.3	2.9
	Lesser whitethroat	1.4	–
	Spotted fly-catcher	2.9	2.9

associated with bank-side scrub and woodland, and also the pied, grey and yellow wagtails were reduced in managed sections.

Mammals

Otter. Of the mammals associated with the river and its banks, the otter is probably of most interest. Although traditionally an 'otter' river, with hunting widely practised in the eighteenth and nineteenth centuries, nevertheless in recent years there has been a decline in numbers comparable with that recorded elsewhere in England and Wales. Between the periods 1950–55 and 1966–71 the Hawkstone Otter Hunt, which has hunted extensively in the area, noted a 55% decrease in its success rate and in 1957 the Wye Otter Hounds had ceased hunting because they were having too many 'blank' days. Despite this decline, recent surveys have shown that otters are still widely distributed within the catchment (Fig. 48).

An extensive survey of the catchment was carried out during the period 1977–79, the Welsh sections being completed in 1977 and 1978 and the English sections in 1979.

The distribution of otters was indicated by signs of recent activity, notably footprints and faeces (spraint), the latter frequently being deposited on prominent features such as rocks, fallen trees and ledges under bridges. About eight sites, each 600 m long, were examined carefully in each 10 km² of catchment. In addition, areas around most bridges between the sampling sites were searched. It was concluded from this survey that otters were present on the R. Wye and many of the tributaries from Glasbury to Rhayader, as well as upper sections of the R. Lugg and R. Arrow, although they were surprisingly absent from the R. Irfon. This absence was temporary and perhaps caused by disturbance associated with weir and bridge building for in 1980 (Crawford, personal communication), after construction had ceased, otters were recorded over much of the length of the R. Irfon.

Further observations in 1980 (Macdonald & Mason, personal communication) confirm the unsuitability of the main river upstream of Rhayader, associated with the lack of cover and perhaps shortage of food, but suggest that otters are present downstream of Glasbury, certainly as far as Hay-on-Wye.

Mink. The mink was certainly established in the catchment by 1968 when three animals were caught. The initial centres of distribution were on the R. Lugg and lower Wye, from Ross to Hereford. By the mid-1970s, after bounties had been introduced for their capture, it was evident that mink were widespread in the lower catchment and had spread to the R. Monnow and R. Trothy. By 1980 signs of mink were found in the R. Irfon and other tributaries of the Wye in the middle and upper reaches (Crawford, personal communication), and it can be assumed that by that date there were few parts of the catchment where it was absent. The highest densities of mink have still remained in the initial centres of distribution and about 100 bounties have been claimed each year, suggesting that in some localities the species is relatively abundant. Surveying those parts of the Wye catchment in England during 1979, Crawford (personal communication) found positive evidence of mink at about 30% of the sites which were thoroughly examined.

Plate 3   Collecting invertebrates with a cylinder sampler in the headwaters at Llangurig (near W8).

7. The future

Changes in the water quality and ecology of the R. Wye seem most likely to result either from changes in land-use within the catchment or from water resource developments such as those proposed for the enlargement of Craig Goch Reservoir and for the regulation of the flow of the Wye, that regulation being with or without further reservoir capacity. In addition, the fishery resources of the catchment, particularly those of salmon, are vulnerable both to the current high rates of cropping at sea and within the river (this latter problem has already been discussed in Chapter 5).

Land-use changes

In lowland areas with mixed arable farming and grassland, further increases in the application of nitrogenous fertiliser are expected, particularly on grassland (Church 1976). Whilst such fertilisation is likely to lead to increased concentrations of nitrate within the Lugg, Monnow and lower Wye, the concentrations already seem too high, averages exceeding 0.5 mg/l NO_3-N in most of the river length, to affect the growth of aquatic plants. Furthermore, increases in nitrate are unlikely to change the composition of plant associations within the river. Other major changes in farming practices within these lowland areas which are likely to influence the river are not envisaged, although an increase in the proportions of both arable land and temporary leys seems likely, and this could lead to increasing silt loads and concentrations of pedologically derived substances and fertilisers (Langbein & Schumm 1958).

The major changes in land-use are likely to occur within the uplands and two directions seem probable, these being afforestation and improvements to sheep pasture (Centre for Agricultural Strategy 1978).

The uplands of the Wye catchment (above about 320 m) comprise only about 1% woodland (Newson 1979), and it has been proposed that over much of upland Britain there are advantages in greatly increasing the area of forests, particularly of conifers. In the Wye catchment, extensive afforestation is likely to lead to substantially reduced yields of water. In the Elan Valley sub-catchment alone, if it is assumed that yields are reduced by 15% (Newson 1979), complete afforestation of the catchment would reduce the yield by 0.9 m³/s, annually worth about £0.3 million as untreated bulk supply at 1980 costs. Clearly in catchments extensively used for water supply, the economics of alternative land-use must recognise the crucial importance of water resources.

Substantial afforestation of the uplands would not only change the yield and pattern of run-off, extending the time of storm run-off (Newson 1979), but would also change water quality. At times of soil disturbance, associated both with planting and cropping, soil erosion on the steep slopes, similar to those in the upper Wye catchment, is commonly substantial and leads to enhanced silt concentrations in upland streams and sedimentation in both river gravels (Gibbons & Salo 1973) and impoundments. Where such planting and early management are accompanied by application of fertilisers, their loss is sometimes appreciable. In impounded sub-catchments, such as the Elan, these changes in water quality could increase the currently low phytoplankton productivity of the reservoirs.

Possibly of more significance are the more permanent changes in water quality frequently associated with afforestation, particularly the acidity and increased concentrations of aluminium (Harriman & Morrison 1981).

Water resource developments

In the review of water resources in England and Wales (Water Resources Board 1973), it was recommended that Craig Goch, the uppermost of the existing series of direct-supply reservoirs in the Elan Valley, be enlarged and used to regulate both the rivers Severn and Wye at times when natural flows were too low to satisfy projected increases in demand on downstream abstractions. In the initial scheme, a reservoir with a storage volume of 534×10^6 m³ was envisaged, collecting water (by gravity and by pumping) from the upper Wye, Severn, Ystwyth and Rheidol as well as from its own catchment, and yielding almost 14 m³/s.

With the subsequent downturn in the projected water demand since 1973, the preferred Craig Goch scheme has contracted such that the proposed storage volume is now reduced to 238×10^6 m³, and the gathering grounds limited to the reservoir's natural catchment and modest sources in the upper Ystwyth which may be transferred by gravity. Thus whilst the Severn would be a recipient of Craig Goch water it would no longer be a donor, and the Wye would receive, almost entirely, water derived from its own catchment – thus avoiding the potential problems, particularly that of the transfer of fish pathogens and other unwanted organisms, associated with intercatchment transfer. Furthermore a reduced regulation scheme (Elan Redeployment Scheme) has already been approved for the Wye which uses existing reservoir storage in the Elan Valley and redeploys current compensation flows to the R. Elan from these reservoirs, reducing such flows to save water for discharge when natural flows are considered inadequate to sustain abstracted uses. It is proposed that the maximum regulation discharge, about

2 m³/s, should be released when the natural flow at Monmouth is less than 14 m³/s plus any abstractions. This compares with a maximum regulation discharge to the Wye of about 6 m³/s projected for the original Craig Goch Scheme.

It is evident from Fig. 49, showing the flows at Monmouth during the summer months of 1975–77, that in 1976, the driest on record, the Wye would have been regulated for substantial periods but that in 1977, a rather wet year, regulation would rarely have been necessary. However in 'normal' years summer flows of about 10 m³/s are not uncommon in the lower Wye, and, at those flows, regulation releases could comprise, even in these downstream reaches, 40% of the total flow with the initial Craig Goch Scheme and 20% with the Elan Redeployment Scheme. Outside the months of June to September natural flows are generally too high to warrant regulation releases.

Such releases may modify the river and its ecology, directly, as a consequence of flow change, or, indirectly, as a consequence of water quality changes on storage or of increasing the proportion of upland waters which are derived from catchments with base-poor soils. The former may be regarded as flow- and the latter as storage-related effects.

Effects of flow changes

Flow changes may exert their effects through changes in velocity, water depth or channel width. In the middle and lower catchments of the Wye flow changes principally result in changing velocity, whereas in the upper catchment both velocity and cross-sectional area are equally affected (Edwards, in press). As a consequence of regulation, the minimum velocities throughout the river downstream of the Elan–Wye confluence would be increased as would be depth and width in the upper reaches (Fig. 50); these changes are small, however, compared with normal variations. Associated with the increased velocity, the maximum residence times of water within the river would be appreciably reduced by regulation.

Within the velocity range influenced by regulation, whilst there may be changes in the erosion and deposition rates of inorganic sediments, of much greater potential significance is the effect of increased velocity on the behaviour of organic particles, which are almost neutrally buoyant. Observations have only been made on drifting invertebrates and during experimental releases of water from Caban Coch, discussed in Chapter 4, those observations have demonstrated that increases in reservoir discharge, comparable with regulation, increase both the total number and density of drifting organisms, although this increase seems only temporary, invertebrates becoming adjusted to the modified flow regime. With increased water velocities causing both increased loss-rates from the benthos, and, at the higher velocities, an extension of distances transported, some enhanced downstream population displacement might be expected; although with normal disper-

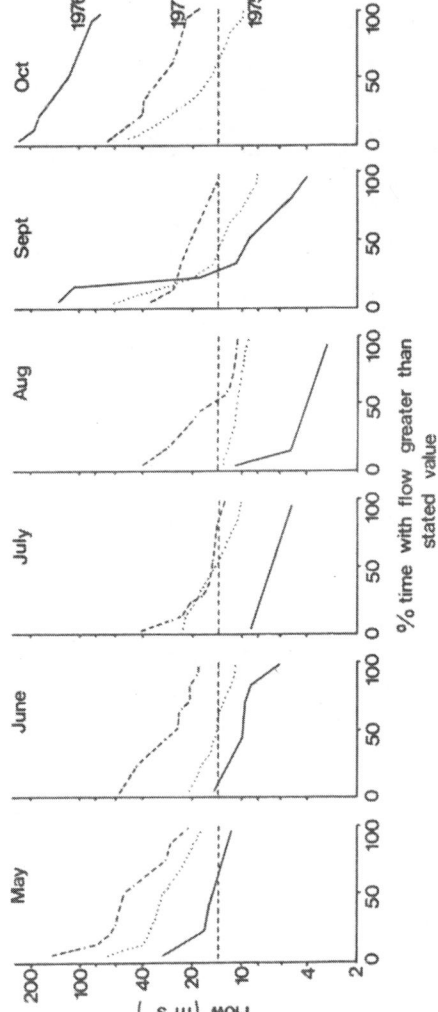

Fig. 49 Distribution of daily river flows at Monmouth for the months of June, July, August and September in the period 1975–77 compared with the flow (14 m³/s) at which regulation would take place.

sive mechanisms, particularly of species with an aerial flight stage, major shifts in the distribution of benthic invertebrates seem unlikely (Hemsworth & Brooker 1979).

Changes in minimum water velocity during the summer are likely to be of major significance in the lower reaches where dense stands of water-buttercup are present, generally until July. The effects of velocity are complex, influencing reaeration, respiratory and photosynthetic rates (Edwards & Crisp, in press). Nevertheless it seems certain that flow supplementation will improve the oxygen status of the river during the critical periods of maximum plant biomass and subsequent plant decay and will reduce the likelihood of mass mortality of salmon such as that which occurred in the drought of 1976 (Brooker et al. 1977).

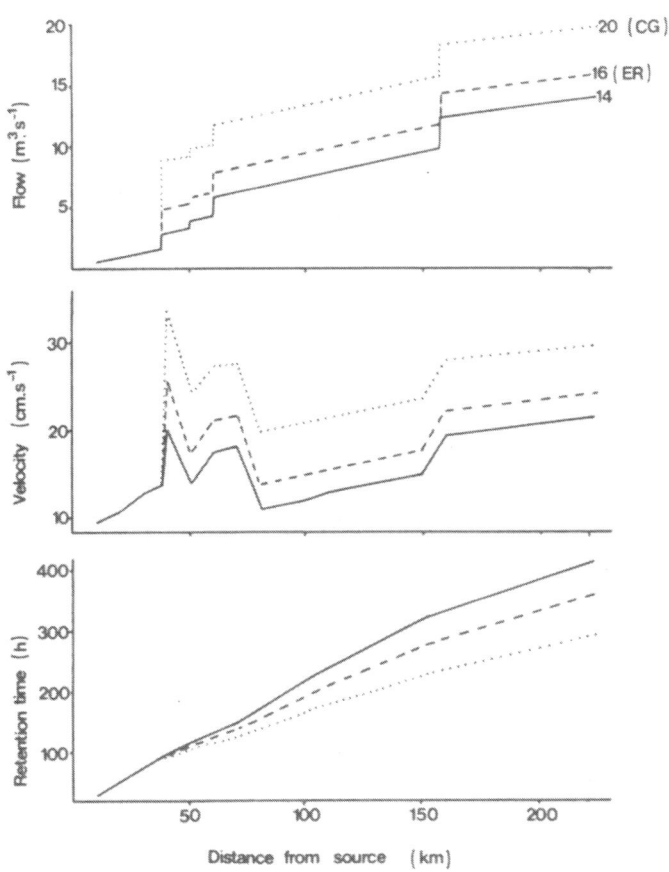

Fig. 50 Effects of two schemes of regulation (see text) on flow, velocity and residence times of the Wye from the source to Monmouth.

Table 22 Simulated mean concentrations (mg/l) and ranges: (10–90 percentile) of alkalinity, chloride, phosphate-P and total hardness under natural flow conditions (N), Elan Redeployment (ER) and Craig Goch (CG) regulation conditions during contrasting 12-month periods (see text).

	Dry period			Wet period		
	N	ER	CG	N	ER	CG
Alkalinity						
Mean	101	89	77	91	88	84
Range	66–141	66–112	65–92	60–130	60–123	60–106
Chloride						
Mean	17.4	16.1	14.7	16.0	15.8	15.3
Range	13.5–23.5	13.5–20.2	13.5–16.5	13.5–20.0	13.5–19.8	13.5–18.3
Phosphate-P						
Mean	0.140	0.119	0.097	0.086	0.083	0.079
Range	0.069–0.263	0.069–0.203	0.069–0.139	0.064–0.133	0.064–0.120	0.064–0.103
Total hardness						
Mean	123	110	97	115	112	107
Range	87–169	88–135	84–112	86–154	86–151	86–131

The effects of regulation on salmon movements and on angling in the Wye are not easy to predict but are not likely to be as substantial as natural freshets in inducing upstream movement of fish because: (1) supplementaty flows are abstracted at Monmouth and do not increase flows in the lowest reaches, (2) supplementation is not pulsed but merely balances abstraction, and (3) the quality of stored water is different from that of natural freshets.

Nevertheless, in the upper reaches where regulation discharges represent a large proportional increase in summer flows, some improvement in salmon capture may occur despite differences from natural freshets with respect to water quality.

The effect of regulation on residence time may be particularly important with respect to the removal of nutrients and the concomitant growth of micro-organisms, particularly algae. It is difficult, however, to predict even the direction of concentration change, for, whilst initial concentrations will be lower with increased dilution from nutrient-poor reservoir sources, loss rates of substances such as SiO_2 and NO_3 could also be reduced with shorter residence times within the river. The concentration of such substances is only of major importance if the river water is subsequently impounded for periods of time sufficient to induce algal growth.

Effects of storage

Using the flow-separation model referred to in Chapter 2 (Oborne 1981), the effects of regulation on the concentration of certain substances, which may be regarded as conservative, at Monmouth have been calculated for the 12-month periods when flows were 38 and 118% of the long-term average. Table 22 shows the results for alkalinity, chloride, phosphate and total hardness, and with all these determinands regulation reduces not only mean concentration, but also ranges, particularly in dry years when regulating releases are most frequent. The ecological consequences are not easy to predict and, with the possible exception of the effects of changes in calcium concentration (Edwards et al. 1978) on crustaceans and molluscs, are not likely to be substantial. The reduction in calcium concentration, at Monmouth in the lower reaches, from 45 to 33 mg/l in the case of the Craig Goch Scheme, could be of more significance to supplies abstracted for industrial use.

In addition to the extensive effects, associated with changes in conservative properties, of releasing stored water, there are potentially several local effects downstream of the dam associated with differences in such properties as the temperature and oxygen content of reservoir and natural river water. Fig. 9 shows the mean monthly water temperatures during the period 1977–79 and the cumulative number of day-degrees at sites in the Elan and in the Wye above and below the Elan confluence. With the current operational procedure of releasing hypolimnetic compensation water to the river, temperatures in the R. Elan are substantially below those in the neighbouring R. Wye, particularly from May to August.

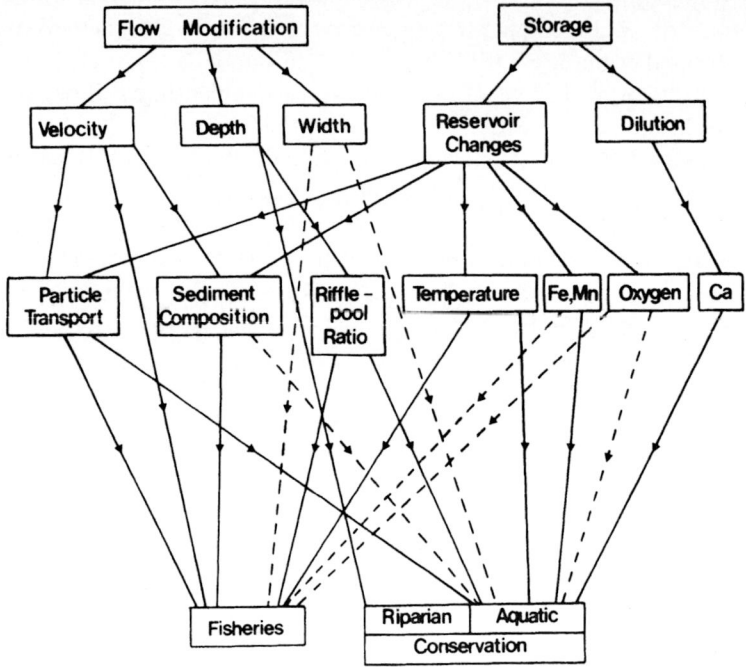

Fig. 51 Potential local effects of regulation, through flow modification and storage, on fisheries and wild-life conservation of the upper Wye. Hatched lines indicate effects least likely to be of major significance.

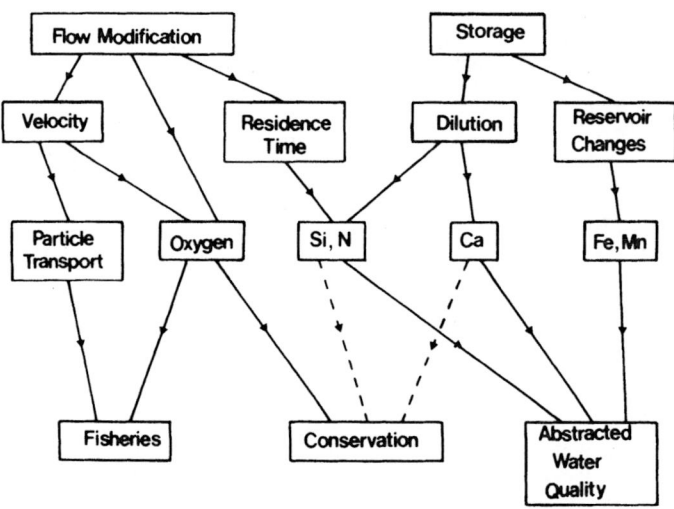

Fig. 52 Potential extensive effects of regulation, through flow modification and storage, on fisheries, wildlife conservation and the quality of abstracted water.

120

Nevertheless, within the R. Elan there is an appreciable increase in temperature from the reservoir to the Wye confluence, and below the confluence the effect of the Elan is not detectable in most years. Even with the increased regulation discharges proposed, it is unlikely that the temperature regimes of extensive reaches will be modified, the river rapidly equilibrating to the heat sources and sinks related to climatological, topographical and geological characteristics of the catchment. The Ithon and Irfon, two major tributaries only 12 and 20 km downstream of the Elan confluence also minimize the extent of temperature modification.

Within such localised areas, however, the reduction of temperature could have ecological consequences. For example, it would seem from studies of the effect of temperature on the growth of brown trout (Elliott 1975) that in the upper Elan the growth rate is likely to be less than 65% of that at sites in the upper Wye both above and below the confluence (Edwards, in press).

The oxygen content of hypolimnetic water in Craig Goch is likely to be partially de-oxygenated, and unless de-stratification procedures are adopted within the reservoir or epilimnetic water is abstracted, then, without artificial re-aeration, localised reaches of the receiving river will probably have reduced oxygen concentrations. Rates of surface re-aeration are rapid, however, in fast, shallow rivers, such as the R. Elan and upper Wye, and the lengths affected are likely to be short.

On the other local effects, the most important, and likely to result from the discharge of phyto- and zoo-plankton from the reservoir, is the development of high densities of filter-feeders such as larvae of the Simuliidae, net-spinning Trichoptera and *Hydra*. Even at the low current compensation flows, it is estimated that 15 t fresh wt/y of zooplankton is discharged from Caban Coch Reservoir. The capricious nature of regulation and the concomitant discharge of particles of organic debris, possibly derived from peat, and rich in iron and manganese, could prevent the development of such filter-feeders. The latter factor is regarded as important in restricting the existing fauna of the R. Elan to one characteristic of rivers polluted with ferric hydroxide (Chapter 4).

General conclusions

The potential impacts of a major reservoir development leading to the regulation of the flow of the R. Wye, both locally at the site of release and more extensively, are summarised in Figs. 51 and 52. Local effects are principally restricted to 'in situ' resources, fisheries and conservation, and, because of the effect of releases on water depth, riparian communities must be included. Of the more extensive effects, the avoidance of low oxygen concentrations in the lower reaches associated with weed-growth and decay could well be most important, particularly in preventing the occasional mortalities of salmon. The softening of the water during summer months could also be significant, particularly with respect to the distribution of organisms and the costs of treatment of water for industrial supply.

Plate 4 Checking the catch of a drift sampler in the river at Newbridge-on-Wye (W18).

122

Appendices

Appendix 1 List of pteridophyte and angiosperm species found on the Wye and four of its major tributaries (Monnow, Lugg, Irfon and Elan).

	Monnow	Lugg	Irfon	Elan	Wye
Pteriodophytae					
Lycopodiaceae					
Lycopodium selago					+
Equisetaceae					
Equisetum arvense	+	+	+		+
E. palustre		+	+	+	
Dennstaedtiaceae					
Pteridium aquilinum					+
Aspidiaceae					
Dryopteris filix-mas					+
Athyriaceae					
Athyrium filix-femina					+
Aspleniaceae					
Asplenium adiantum-nigrum					+
A. trichomanes					+
Phyllitis scolopendrium					+
Blechnaceae					
Blechnum spicant					+
Angiospermae					
Ranunculaceae					
Anemone nemorosa					+
Caltha palustre	+	+	+	+	+
Clematis vitalba					+
Ranunculus acris					+
R. bulbosus					+
R. ficaria					+
R. flammula		+	+	+	+
R. hederaceus		+		+	
R. omiophyllus					+
R. penicillatus	+	+	+		+
R. repens					+
R. sceleratus	+	+			
Thalictrum flavum					+
Trollius europaeus					+
Nymphaceae					
Nuphar alba		+			
Fumariaceae					
Fumaria muralis					+

125

Appendix 1 (continued).

	Monnow	Lugg	Irfon	Elan	Wye
Cruciferae					
Alliaria petiolata					+
Armoracia rusticana					+
Barbarea intermedia					+
B. vulgaris					+
Brassica napus					+
B. nigra					+
B. rapa					+
Capsella bursa-pastoris					+
Cardamine flexuosa					+
Cardamine hirsuta					+
C. impatiens					+
C. pratensis					+
Coronopus didymus					+
Diplotaxis muralis					+
Rorippa amphibia		+			
R. islandica					+
R. nasturtium-aquaticum		+			+
R. palustris	+	+			+
R. sylvestris	+	+			+
Sinapsis arvensis					+
Sisymbrium officinale					+
Resedaceae					
Reseda lutea					+
R. luteola					+
Polygalaceae					
Polygala serpyllifolia					+
P. vulgaris					+
Violaceae					
Viola odorata					+
V. riviniana					+
V. palustris			+	+	+
Caryophyllaceae					
Cerastium glomeratum					+
C. holosteoides					+
Lychnis flos-cuculi					+
Moehringia trinervia					+
Myosoton aquaticum					+
Sagina procumbens	+	+			+
Saponaria officinalis					+
Silene alba					+
Silena alba × dioica					+
S. doica					+
Spergula arvensis					+
Stellaria alsine		+		+	+

126

	Monnow	Lugg	Irfon	Elan	Wye
S. graminea					+
S. media					+
Portulaceae					
Montia fontana		+	+	+	+
M. sibirica			+		+
Hypericaceae					
Hypericum androsaemum					+
H. × desetangsii					+
H. elodes					+
H. perforatum					+
H. pulchrum					+
H. tetrapterum					+
Malvaceae					
Malva sylvestris					+
Geraniaceae					
Geranium pratense					+
G. robertianum					+
Oxalidaceae					
Oxalis acetosella					+
Balsaminacea					
Impatiens glandulifera	+				+
Aquifoliaceae					
Ilex aquifolium					+
Celastraceae					
Euonymus europaeus					+
Aceraceae					
Acer pseudoplantanus					+
A. campestre					+
Papilionaceae					
Genista tinctoria					+
Lathyrus pratensis					+
L. syvestris					+
Lotus corniculatus					+
L. pedunculatus		+	+	+	+
Mediago lupulina					+
Ononis repens					+
O. spinosa					+
Sarothamnus scoparius					+
Trifolium ornithopodioides					+
T. pratense					+
T. repens					+
Ulex europaeus					+

127

	Monnow	Lugg	Irfon	Elan	Wye
Viccia cracca					+
V. hirsuta					+
V. lathyroides					+
V. orobus					+
V. sepium					+
V. sylvatica					
Rosaceae					
Aphanes arvensis					+
Crataegus monogyna					+
Filipendula ulmara	+	+	+		+
Fragaria vesca					+
Geum urbanum					+
Malus sylvestris					+
Potentilla anserina					+
P. erecta		+	+	+	+
P. reptans					+
P. rupestris					+
P. sterilis					+
Prunus padus					+
P. spinosa					+
Rosa arvensis					+
R. canina					+
R. sherardii					+
R. stylosa					+
R. tomentosa					+
R. villosa					+
Rubus fruticosa	+	+			+
R. idaeus					+
Sanguisorba officinalis					+
Sorbus aucuparia					+
Saxifragaceae					
Chrysosplenium oppositifolium	+	+			+
Saxifraga granulata					+
S. stellaris					+
Grossulariaceae					
Ribes nigrum					+
R. rubrum		+			
R. sylvestre					+
Crassulaceae					
Sedum anglicum					+
S. rosea					+
Droseraceae					
Drosera rotundifolia					+

128

	Monnow	Lugg	Irfon	Elan	Wye
Haloragaceae					
Myriophyllum alternifolium			+	+	
M. spicatum	+	+			+
M. verticillatum					+
Callitrichaceae					
Callitriche intermedia hamulata					+
C. obtusangula		+			
C. stagnalis		+			
Lythraceae					
Lythrum salicaria		+			+
Peplis portula					+
Onagraceae					
Chamaemerion angustifolium					+
Circaea lutetiana					+
Epilobium hirsutum	+	+	+		+
E. montanum					+
E. nerterioides					+
E. palustra					+
E. tetragonium					+
Umbelliferae					
Aegopodium podagraria					+
Angelica sylvestris	+	+	+		+
Anthriscus sylvestris					+
Apium graveoleus					+
A. nodiflorum	+				
Conopodium majus					+
Oenanthe crocata	+	+	+	+	+
Heracleum sphondylium					+
Sanicula europaea					+
Torilis japonica					+
Araliaceae					
Hedera helix					+
Cornaceae					
Thelycrania sanguinea					+
Caprifoliaceae					
Lonicerca periclymemum					+
Sambucus nigra					+
Viburnum opulus					+
Rubiaceae					
Galium aparine					+
G. cruciata					+
G. mollugo					+
G. odoratum					+

	Monnow	Lugg	Irfon	Elan	Wye
G. palustre			+		+
G. saxatile					+
G. uliginosum					+
Valerianaceae					
Valeriana dioica	+		+		
V. officinalis					+
Dipsaceae					
Dipsacus fullonum		+			
Succisa pratensis					+
Compositae					
Achillea millifolium					+
A. ptarmica		+	+	+	+
Anthemis cotula					+
Arctium lappa					+
A. minus					+
Artemesia absinthium					+
A. vulgaris					+
Bellis perennis					+
Bidens cernua	+				
B. tripartita	+				+
Centaurea nigra					+
Chamaemelium nobile					+
Chrysanthemum leucanthemum					+
C. parthenium					+
C. vulgare					+
Cirsium arvense					+
C. palustre					+
C. vulgare					+
C. capillaris					+
Eupatorium cannabinum					+
Gnaphalium uliginosum					+
Hieracium pilosella					+
Hypochoeris radicata					+
Lapsana communis					+
Leontodon hispidus					+
Matricaria matricarioides					+
M. recutita					+
Mycelis muralis					+
Petasites hybridus	+	+			
Picris hieracioides					+
Senecio aquaticus					+
S. jacobaea					+
S. vulgaris					+
Solidago virgaurea					+
Sonchus arvensis					+
S. asper					+

	Monnow	Lugg	Irfon	Elan	Wye
S. oleraceus					+
Taraxacum officinale					+
Tussilago farfara	+	+	+		+
Tragoporon pratensis					+
Campanulaceae					
Campanula latifolia					+
C. rotundifolia					+
Wahlenbergia hederacea					+
Ericaceae					
Calluna vulgaris					+
Rhododendron ponticum					+
Vaccinium myrtillus					+
Empetraceae					
Empetrum nigrum					+
Primulaceae					
Anagallis arvensis					+
Lysmachia nemorum	+	+		+	+
L. nummularia					+
L. vulgaris	+	+	+		+
Primula vulgaris					+
Oleaceae					
Fraxinus excelsior					+
Gentianaceae					
Centaurum erythraea					+
Boraginaceae					
Myosotis arvensis					+
M. scorpioides	+	+	+	+	+
M. secunda					+
M. sylvatica					+
Symphytum officinale	+	+			+
Convolvulaceae					
Calystegia sepium					+
Convolvulus arvensis					+
Solanaceae					
Solanum dulcamara	+	+	+		+
Scophulariaceae					
Digitalis purpurea					+
Euphrasia hirtella					+
Linaria vulgaris					+
Melampyrum pratense					+
Mimulus guttatus		+			+
M. guttatus × luteus		+	+		

131

Appendix 1 (continued).

	Monnow	Lugg	Irfon	Elan	Wye
M. moschatus					+
Pedicularis sylvatica					+
Rhinanthus minor					+
Scrophularia auriculata	+	+			
S. nodosa					+
Veronica angallis-aquatica		+			
V. beccabunga	+	+			+
V. catenata	+	+			
V. filiformis					+
V. chamaedrys					+
V. montana					+
V. officinalis					+
V. scutellata	+	+			

Lentibulariaceae
| *Pinguicula vulgaris* | | | | | + |

Labiatae
Ajuga reptans					+
Bettonica officinalis					+
Galeobdolon luteum					+
Galeopsis tetrahit					+
Glechoma hederacea					+
Lamium album					+
L. purpureum					+
Lyocopus europaeus	+	+	+		+
Mentha aquatica	+	+	+	+	+
M. rotundifolia					+
M. × verticillata					+
Origanum vulgare					+
Prunella vulgaris					+
Stachys palustris	+	+	+		
S. vulgaris					+
Teucrium scorodonia					+
Thymus drucei					+

Plantaginaceae
Littorella uniflora			+		+
Plantago lanceolata					+
P. major					+
P. maritima		+			

Polygonaceae
Polygonum amphibium	+	+			+
P. aviculare					+
P. cuspidatum					+
P. hydropiper					+
P. lapathifolium					+
P. persicaria					+

132

	Monnow	Lugg	Irfon	Elan	Wye
Rumex acetosa					+
R. acetosella					+
R. conglomeratus					+
R. crispus					+
R. obtusifolius					+
R. sanguineus					+
Euphorbiaceae					
Mercurialis perennis					+
Ulmaceae					
Ulmus glabra					+
U. procera					+
Cannabinaceae					
Humulus lupulus					+
Urticaceae					
Urtica dioica					+
U. urens					+
Betulaceae					
Alnus glutinosa					+
Betula pendula					+
B. pubescens					+
Corylaceae					
Carpinus betulus					+
Corylus avellana					+
Fagaceae					
Fagus sylvaticus					+
Quercus petraea					+
Q. robur					+
Salicaceae					
Populus tremula					+
P. × euramuric					+
Salix alba					+
S. aurita					+
S. caprea					+
S. cinerea					+
S. decipiens					+
S. fragilis					+
S. triandra					+
S. viminalis					+
Alismataceae					
Alisma plantago-aquatica	+	+	+		+
Sagittaria sagittifolia		+			

	Monnow	Lugg	Irfon	Elan	Wye
Butomaceae					
Butomus umbellatus		+			
Hydrocharitaceae					
Elodea canadensis		+			+
Orchidaceae					
Dactylorchis maculata					+
Iridaceae					
Iris pseudacorus				+	+
Liliaceae					
Allium schoenoprasum					+
A. ursinum					+
A. vineale					+
Endymion non-scriptus					+
Narthecium ossifragum			+		+
Polygonatum multiflorum					+
Juncaceae					
Juncus acutiflorus	+	+	+	+	+
J. articulatus			+	+	+
J. bufonius	+		+		+
J. bulbosus		+		+	
J. effusus		+	+	+	+
J. inflexus	+	+			+
J. kochii					+
J. squarrosus					+
Luzula campestris					+
L. multiflora					+
L. sylvatica					+
Typhaceae					
Typha latifolia		+			
Sparganiaceae					
Sparganum erectum					+
Araceae					
Arum maculatum					+
Lemnaceae					
Lemna minor		+			+
Potamogetonaceae					
Potamogeton crispus	+	+	+		+
P. lucens					+
P. pectinatus		+			+
P. perfoliatus	+	+			+
P. polygonifolius				+	+
P. × salicifolius					+

134

	Monnow	Lugg	Irfon	Elan	Wye
Zannichelliaceae					
Zannichellia palustris	+	+			
Cyperaceae					
Carex acuta					+
C. curta			+	+	+
C. demissa		+		+	+
C. echinata				+	+
C. hirta	+	+	+		
C. laevigata					+
C. ovalis		+	+	+	+
C. nigra		+	+	+	+
C. panicea		+			+
C. pseudocyperus					+
C. remota	+	+	+	+	+
C. rostrata				+	+
Eleocharis palustris	+		+	+	+
Eriophorum angustifolium					+
Scirpus lacustris		+			
S. sylvaticus	+	+	+		
S. emersum		+			
S. erectum	+	+	+		
Gramineae					
Agropyron caninum					+
A. repens					+
Agrostis stolonifera	+	+	+	+	+
A. tenius					+
Aira caryophyllea					+
Alopecurus geniculatus	+	+			+
Anthoxanthum odoratum		+	+	+	+
Arrhenatherum elatius					+
Brachypodium sylvaticum					+
Bromus sterilis					+
Cynosurus cristatus					+
Dactylis glomerata					+
Deschampsia caespitosa	+	+	+	+	+
D. flexuosa					+
Festuca arundinacea					+
F. gigantea					+
F. ovina					+
F. rubra					+
F. vivipara					+
Glyceria declinata		+			+
G. fluitans	+	+	+	+	+
G. plicata	+				+
G. pedicillata	+				
Holcus lanatus					+

135

Appendix 1 (continued).

	Monnow	Lugg	Irfon	Elan	Wye
H. mollis					+
Lolium perenne					+
Melica uniflora					+
Molinia caerculea				+	+
Nardus stricta				+	+
Phalaris arundinacea	+	+	+	+	+
Phleum bertolonii					+
P. pratense					+
Phragmites australis		+	+		
Poa annua					+
P. nemoralis					+
P. trivialis					+
Sieglingia decumbens					+
Vulpia bromoides					+

Appendix 2 List of lichen and bryophyte species found in the Wye and four of its major tributaries
(Monnow, Lugg, Irfon and Elan).

	Monnow	Lugg	Irfon	Elan	Wye
Lichens					
Dermatocarpon fluviatile					+
Peltigera sp.					+
Verrucaria praetermissa	+	+	+		
Bryophytes					
Sphagnales					
Sphagnum auriculatum					+
S. girgensohnii					+
S. palustre					+
S. papillosum					+
S. plumulosum					+
S. recurvum					+
S. squarrosum					+
Andreaeales					
Andreaea rothii					+
Polytrichales					
Atrichum crispum					+
A. undulatum					+
Oligotrichum hercynium			+	+	+
Polytrichum aloides					+
P. commune			+	+	+
P. formosum					+
P. piliferum	+				+
P. urnigerum					+
Fissidentales					
Fissidens adianthoides		+			
F. bryoides					+
F. crassipes	+	+			+
F. cristatus					+
F. taxifolius					+
F. viridulas	+	+			
Octodiceras fontanum	+				
Dicranales					
Blindia acuta		+			
Campylopus atrovirens				+	+
C. flexuosus					+
C. subulatus					+
Ceratodon purpureus					+
Dichodontium flavescens	+				
D. pellucidum	+				+
Dicranella heteromalla			+		+
D. schreberama					+

137

	Monnow	Lugg	Irfon	Elan	Wye
D. squarrosus					+
D. varia			+		+
Dicranum majus					+
D. scoparium					+
Ditrichum heteromallum					+
Pleuridium subulatum					+
Pottiales					
Barbula cylindrica	+				+
B. recurvirostre	+	+			
B. unguiculata					+
Cinclidotus fontinaloides	+	+	+		+
Trichostomum brachydontium					+
T. sinuosum					+
Grimmiales					
Grimmia alpicola	+		+	+	
G. doniana					+
G. trichophylla					+
Rhacomitrium aciculare			+	+	+
R. aquaticum				+	+
R. canescens				+	+
R. fasciculare					+
R. heterostichum					+
R. lanuginosum					+
Funariales					
Funaria hygrometrica		+			
Physocomitrium pyriforme					+
Eubryales					
Breutelia chrysocoma					+
Bryum alpinum				+	+
B. bicolor		+			
B. caespiticium					+
B. capillare					+
B. pseudotriquetrum					+
B. rubens					+
Mnium hornum	+	+	+	+	+
M. punctatum	+	+	+	+	+
M. rostratum	+	+			
M. undulatum			+		+
Philonotis calcarea					+
P. fontana	+	+	+		+
Pohlia annotina					+
P. carnea	+	+			
P. deliculata					+
P. elongata				+	
P. proligera			+		

138

Appendix 2 (continued).

	Monnow	Lugg	Irfon	Elan	Wye
Isobryales					
Climacium dendroides					+
Fontinalis antipyretica	+	+	+	+	+
F. squamosa			+	+	+
Omalia trichomanoides					+
Orthotrichum rivulare	+				
Ptychomitrium polyphyllum					+
Thamnium alopecurvum	+	+	+		+
Hypnobryales					
Acrocladium cuspidatum		+	+		+
A. stramineum					+
Amblystegium fluviatile	+	+	+		+
A. riparium	+	+	+		
A. tenax					+
Brachythecium albicans		+			
B. plumosus	+			+	+
B. rivulare	+	+			+
B. rutabulum	+	+			+
B. velutinum					+
Cirriphyllum crassinervium					+
Cratoneuron commutatum		+			
C. filicinum	+	+	+		
Drepanocladus exannulatus					+
D. fluitans			+		+
D. uncinatus					+
Eurhynchium praelongum				+	+
E. unparioides	+	+	+	+	+
E. swartzii				+	
Heterocladium heteropterum				+	
Hygrohypnum luridum		+			
H. ochraceum		+	+	+	+
Hylocomium flagellare					+
H. splendens		+	+	+	+
Hyocomium armoricum		+	+	+	
H. umbratum				+	
Hypnum cupressiforme					+
Isothecium myosuroides					+
I. myurum					+
Plagiothecium curvifolium					+
Pleurozium schreberi					+
Thuidium philibertii				+	+
T. tamariscinum					+
Rhytidiadelphus loreus					+
R. squarrosus			+		

Appendix 2 (continued).

	Monnow	Lugg	Irfon	Elan	Wye
Marchantiales					
Conocephalui conicum	+	+	+		+
Lunularia cruciata		+			+
Marchantia polymorpha	+	+	+		+
Metzgeriales					
Pellia endiviifolia	+	+	+	+	
P. epiphylla		+	+	+	+
Riccardia pingius					+
Jungermanniales					
Barbilophozia floerkei					+
Cephalozia bicuspidata				+	
Chiloscyphus polyanthos					+
Diplophyllum albicans					+
Gymnocolea inflata					+
Lejeunea lamacerina					+
Lophocolea bidentata					+
Marsupella emarginata			+	+	+
Nardia compressa			+	+	+
N. scalaris					+
Odontoschisma sphagni					+
Plagiochyla aspenoides					+
Saccogyna viticulosa					+
Scapania undulata			+	+	+
Solenostoma cordifolium				+	
S. crenulatum					+
S. trista	+	+	+	+	+

Appendix 3   The occurrence of macroinvertebrates in the Wye catchment.

	W4	W8	W13	W17	E6	W18	W24	Ir 1	W29	W35	W43	W53	W55	W61
Coelenterata														
Hydridae						+			+		+		+	
Nematoda														
Nematoda indet.			+		+	+			+		+		+	
Platyhelminthes														
Planaria torva (Muller)		+	+	+					+		+	+	+	+
Polycelis felina (Dalyell)		+	+										+	+
Dugesia lugubris (Schmidt)		+	+	+					+			+	+	+
Phagocata vitta (Duges)	+	+	+	+	+	+	+	+	+					
Dendrocoeleum lacteum (Muller)														+
Oligochaeta														
Nais communis Piguet	+						+						+	
N. variabilis Piguet	+			+										+
N. alpina Sperber	+	+	+	+	+	+	+	+	+		+		+	+
N. elinguis Muller			+	+		+	+	+	+	+			+	+
Pristina idrensis Sperber											+	+	+	+
Stylaria lacustris (L.)									+	+		+	+	+
Uncinais uncinata (Orsted)									+		+		+	+
Psammoryctides barbatus (Grube)								+	+		+	+	+	+
Limnodrilus hoffmeisteri Claparede						+	+	+	+	+	+	+	+	+
Limnodrilus sp.									+	+			+	+
Pelescolex ferox (Eisen)		+	+		+	+		+	+	+	+	+	+	+
Rhyacodrilus coccineus (Vejdovsky)		+	+	+	+	+	+	+	+	+	+	+	+	+
Aulodrilus pluriseta Piguet							+	+	+	+	+	+	+	+
Enchytraeidae	+	+	+	+	+	+	+	+	+	+	+	+	+	+
Lumbriculus variegatus (Muller)	+	+	+	+	+	+	+	+	+	+	+	+	+	+
Stylodrilus heringianus Claparede		+	+	+	+	+	+	+	+	+	+	+	+	+
Eclipidrilus (lacustris?)								+						
Eiseniella tetraedra (Savigny)	+	+	+		+	+	+		+	+	+	+	+	+

141

Appendix 3 (continued).

	W4	W8	W13	W17	E6	W18	W24	Ir 1	W29	W35	W43	W53	W55	W61
Hirudinea														
Piscicola geometra (L.)													+	+
Glossiphonia complanata (L.)				+			+				+	+	+	+
Helobdella stagnalis (L.)											+	+	+	+
Theromyzon tessulatum (Muller)														+
Erpobdella octoculata (L.)			+		+	+	+	+	+		+	+	+	+
Crustacea														
Daphnia sp.													+	
Eurycercus lamellatus Muller					+									
Herpetocypris reptans (Baird)											+	+	+	
Macrocyclops albidus (Jurine)													+	+
Malacostraca														
Asellus aquaticus (L.)								+	+	+	+	+	+	+
Gammarus pulex (L.)									+	+	+	+	+	+
Plecoptera														
Rhabdiopteryx acuminata Klapalek		+												
Brachyptera risi (Morton)		+	+											
Protonemura meyeri (Pictet)	+	+	+	+		+	+	+	+	+	+	+	+	
Amphinemura sulcicollis (Stephens)	+	+	+		+	+	+	+	+	+	+	+		
Nemoura cambrica (Stephens)			+											
Nemoura sp.	+													
Leuctra geniculata (Stephens)	+	+		+	+	+	+	+	+	+	+	+	+	+
L. inermis Keipny					+	+	+	+	+	+	+	+		
L. nigra (Olivier)			+											
L. fusca (L.)	+	+	+	+	+	+	+	+	+	+	+			
Leuctra sp.											+			
Perlodes microcephala (Pictet)		+	+	+	+	+	+	+	+	+	+			
Isoperla grammatica (Poda)	+	+	+	+	+	+	+	+	+	+	+			
Dinocras cephalotes (Curtis)										+				
Perla bipunctata Pictet			+			+	+	+	+					
Chloroperla torrentium (Pictet)	+	+	+	+	+	+	+	+	+	+				
C. tripunctata (Scopoli)	+	+	+	+	+	+	+	+	+	+				

142

Ephemeroptera

Baetis scambus Eaton
B. vernus Curtis
B. buceratus Eaton
B. rhodani Pictet
B. muticus (L.)
B. niger (L.)
Procloeon pseudorufulum Kimmins
Centroptilum luteolum (Muller)
Rhithrogena semicolorata (Curtis)
Heptagenia sulphurea (Muller)
Ecdyonurus venosus (Fabricius)
E. torrentis Kimmins
E. dispar (Curtis)
Paraleptophlebia cincta (Retzius)
Ephemerella ignita (Poda)
Potamanthus luteus (L.)
Ephemera danica Muller
Caenis macrura Stephens
C. moesta Bengtsson
C. rivulorum Eaton

Trichoptera

Rhyacophila dorsalis (Curtis)
Glossosoma conformis Neboiss
Plectrocnemia conspersa (Curtis)
P. geniculata McLachlan
Polycentropus flavomaculatus (Pictet)
Phychomyia pusilla (Fabricius)
Hydropsyche pellucidula (Curtis)
H. angustipennis (Curtis)
H. contubernalis McLachlan
H. siltalai Dohler
Cheumatopsyche lepida (Pictet)
Hydroptilidae
Potamophylax (cingulatus?)
Stenophylax (vibex?)
Beraeidae

143

	W4	W8	W13	W17	E6	W18	W24	Ir 1	W29	W35	W43	W53	W55	W61
Athripsodes (albifrons?)	+							+	+	+	+	+	+	+
Mystacides (azurea?)														+
Leptoceridae				+										
Lepidostoma hirtum (Fabricius)		+		+	+		+	+		+			+	+
Brachycentrus subnubilus Curtis					+		+	+	+	+				
Sericostoma personatum (Spence)	+	+	+	+	+	+		+						+
Coleoptera														
Haliplidae												+	+	+
Oreodytes rivalis Gyllenhal														+
O. septentrionalis Gyllenhal													+	
Hydroporus elegans Panzer														+
Dytiscidae larvae		+	+									+	+	+
Gyrinidae larvae		+	+		+	+	+	+	+	+	+	+	+	+
Hydraena gracilis Germar		+	+	+	+	+	+	+	+	+	+		+	+
Cyphon sp.?		+	+											
Elmis aenea (Muller)	+	+	+	+		+	+	+	+	+	+	+	+	+
Esolus parallelepipedus (Muller)	+	+	+	+	+	+	+	+	+	+	+	+	+	+
Limnius volckmari (Panzer)	+	+	+	+	+	+	+	+	+	+	+	+	+	+
Oulimnius tuberculatus (Muller)	+	+	+	+	+	+	+	+	+	+	+	+		
Normandia nitens (Muller)												+		
Riolus sp.								+					+	+
Curculionidae														
Coleoptera spp.						+								
Megaloptera														
Sialis lutaria (L.)											+	+	+	+
S. fuliginosa Pictet									+		+	+	+	+
Odonata														
Coenagriidae larvae												+		
Hemiptera														
Aphelocheirus aestivalis (Fabricius)									+	+	+	+	+	+
Sigara dorsalis (Leach)														+
Corixidae larvae													+	+

Diptera (various)

Tipula sp.

Dicranota sp.

Ceratopogonidae

Hemerodromia sp.

Atherix ibis (Fabricius)

Tabanidae

Ephydra sp.

Psychodidae

Diptera larvae

Diptera (Chironomidae)

Macropelopia (nebulosa?)

Procladius (choreus?)

Ablabesmyia (longistyla?)

Thienemannimyia spp.

Trissopelopia (longimana?)

Zavrelimyia sp.

Diamesa (insignipes?)

Pothastia gaedii (Meigen)

P. longimana Kieffer

Brillia modesta (Meigan)

Cricotopus bicinctus (Meigen)

C. (trifascia?)

Cricotopus (8 spp.)

Isocladius 'sylvestris' spp.

Eukiefferiella clypeata (Kieffer)

E. (discoloripes?)

E. (verralli?)

E. claripennis (Lundbeck)

E. brevicalcar

E. calvescens (Edwards)

E. minor (Edwards)

E. ikleyensis/devonica

Eukiefferiella sp. 9

Heterotrissocladius (marcidus?)

Microcricotopus rectinervis (Kieffer)

Prodiamesa olivacea (Meigen)

Appendix 3 (continued).

	W4	W8	W13	W17	E6	W18	W24	Ir 1	W29	W35	W43	W53	W55	W61
Psectrocladius (psilopterus?)	+	+	+		+	+	+	+	+	+		+	+	+
Rheocricotopus spp.		+	+	+	+	+	+	+	+	+	+	+	+	+
Synorthocladius semivirens (Kieffer)		+	+	+		+	+	+	+	+	+	+	+	+
Corynoneura (lacustris?)	+	+	+	+	+	+	+	+	+	+	+	+	+	+
Metriocnemus sp.		+	+	+		+				+	+			
Thienemanniella (vittata?)	+	+	+	+								+	+	+
Thienemanniella sp. 2	+	+	+	+	+	+	+	+	+	+	+	+	+	+
Orthocladiinae (11 spp.)				+		+	+	+	+	+	+	+	+	+
Cryptochironomus sp.									+	+	+		+	+
Demicryptochironomus vulneratus (Zett.)													+	+
Endochironomus sp.?		+	+		+	+			+	+	+		+	+
Glyptotendipes sp.?										+			+	+
Microtendipes (chloris?)			+	+										
M. (rydalensis?)						+	+		+	+	+		+	
M. sp.					+									
Parachironomus 'varus' spp.				+	+									
Paralauterborniella sp.?		+		+		+		+	+	+	+	+	+	+
Paratendipes (albimanus?)								+	+	+	+	+	+	+
Phaenopsectra sp.?						+					+		+	+
Polypedilum 'convictum' spp.						+	+	+	+	+	+	+	+	+
P. 'nubeculosum' sp. 1		+	+	+	+	+	+	+	+	+	+	+	+	+
P. 'nubeculosum' sp. 2		+	+			+	+	+	+	+	+	+	+	+
P. (laetum?)														
Xenochironomus xenolabis (Kieffer)			+	+	+	+	+	+	+	+	+	+	+	+
Chironomini sp.										+	+	+	+	+
Cladotanytarsus (vanderwulpi?)	+	+	+	+	+	+	+	+	+	+	+	+	+	+
Micropsectra (4 spp.)	+													
Tanytarsus sp.	+	+	+	+	+	+	+	+	+	+	+	+	+	+
Rheotanytarsus sp. 1		+	+	+	+	+	+	+	+	+	+	+	+	+
Rheotanytarsus sp. 2		+											+	
Stempellina bausei (Kieffer)						+								

Diptera (Simuliidae)

Taxon														
Eusimulium latipes Meigen	+	+	+	+	+	+	+	+	+	+	+	+	+	
E. brevicaule Dorier et Grenier	+	+	+	+	+	+	+	+	+	+	+	+	+	+
E. 'aureum' spp.	+	+	+		+	+		+	+	+	+	+	+	+
Wilhelmia salopiense Edwards				.			+		+		+	+	+	+
W. 'equinum' spp.	+	+	+	+	+	+	+	+	+	+	+	+	+	+
Simulium reptans L.	+	+	+	+	+	+	+	+	+	+	+	+	+	+
S. reptans car. *galeratum*	+								+	+		+	+	
S. erythrocephalum Degeer	+	+	+	+	+	+	+	+	+			+	+	+
S. monticola Freiderichs	+	+		+	+	+	+	+	+	+	+	+	+	
S. variegatum Meigen	+	+	+	+	+	+	+	+	+	+	+			
S. ornatum Meigen	+	+	+	+		+	+			+		+	+	
S. nitidifrons Edwards	+	+	+	+	+	+	+	+	+	+				
Simulium sp.				+	+	+	+	+	+	+	+	+	+	+

Arachnida

Taxon														
Hydracarina	+	+	+	+	+	+	+	+	+	+	+	+	+	+

Mollusca

Taxon														
Theodoxus fluviatilis (L.)		+										+	+	+
Valvata macrostoma Morch									+	+	+	+	+	+
Potamopyrgus jenkinsi (Smith)						+	+		+		+	+	+	+
Bythynia (leachi?)												+	+	+
Lymnaea peregra (Muller)					+					+	+	+	+	+
Physa fontinalis (L.)										+				+
Planorbis albus Muller													+	+
P. laevis Alder				+	+								+	+
Ancylus fluviatilis Muller	+		+		+	+	+	+	+	+	+	+	+	+
Sphaerium corneum (L.)		+	+			+		+	+	+	+	+	+	+
Pisidium nitidum Jenyns												+	+	+

147

Appendix 4 Flight periods (+) of adult insects collected at Newbridge-on-Wye 1976–78.

	J	F	M	A	M	J	J	A	S	O	N	D
Plecoptera												
Rhabdiopteryx acuminata Klapelek			+	+	+							
Brachyptera risi (Morton)			+	+	+							
Protonemura meyeri (Pictet)			+	+	+							
Amphinemura sulcicollis (Stephens)					+							
Leuctra geniculata (Stephens)								+	+			
L. inermis Kempny					+							
L. fusca (L.)									+	+	+	+
Isoperla grammatica (Poda)					+	+						
Chloroperla torrentium (Pictet)							+					
C. tripunctata (Scopoli)					+							
Ephemeroptera												
Baetis scambus Eaton								+				
B. vernus Curtis								+	+			
B. rhodani (Pictet)				+		+	+	+	+		+	
B. indet.			+	+	+	+	+					
Ecdyonurus dispar (Curtis)							+					
Ephemerella ignita (Poda)						+	+	+	+			
Caenis rivulorum Eaton						+						
Rhithrogena semicolorata (Curtis)						+						
Trichoptera												
Rhyacophila dorsalis (Curtis)				+	+	+	+	+	+	+	+	
R. munda McLachlan									+	+		+
Glossosoma conformis Neboiss				+	+	+	+	+	+	+	+	
G. intermedium (Klapelek)									+			
Agapetus fuscipes Curtis					+							
A. ochripes Curtis						+	+		+			
A. delicatulus McLachlan							+					
Plectrocnemia geniculata McLachlan								+	+			
Polycentropus flavomaculatus (Pictet)						+	+	+				
Tinodes waeneri (L.)						+	+					
Psychomyia pusilla (Fabricius)						+	+	+				
Hydropsyche pellucidula (Curtis)					+	+						
H. siltalai (Dohler)							+					
H. sp. indet.						+	+	+	+			
Cheumatopsyche lepida (Pictet)							+					
Hydroptila sparsa Curtis						+		+	+			
H. simulans Mosely						+	+	+	+			
H. occulta (Eaton)							+					
H. tineoides Dalman							+					
H. forcipata (Eaton)								+	+	+	+	
Oxyethira indet.								+	+			
Hydroptilidae indet.					+	+		+	+			
Ecclisopteryx guttulata (Pictet)												
Anabolia nervosa Curtis										+		
Potamophylax latipennis (Curtis)							+	+	+			

148

Appendix 4 (continued).

	J	F	M	A	M	J	J	A	S	O	N	D
Halesus digitatus (Schrank)									+			
Arthripsodes albifrons (L.)							+					
Mystacides azurea (L.)									+			
Oecetis testacea (Curtis)							+					
Goera pilosa (Fabricius)							+	+				
Lepidostoma hirtum (Fabricius)							+	+				
Sericostoma personatum (Spence)						+	+					
Megaloptera												
Sialis lutaria (L.)					+							
Diptera (Chironomidae)												
Procladius choreus (Meigen)							+					
Ablabesmyia longistyla Fittkau							+	+	+			
Conchapelopia pallidula (Meigen)							+					
Nilotanypus dubius (Meigen)							+	+				
Rheopelopia eximia (Edwards)							+	+				
Thienemannimyia (woodi?)							+		+			
Trissopelopia longimana (Staeger)								+	+			
Diamesa tonsa (Walker)			+									
Potthastia longimana Kieffer					+							
P. gaedii (Meigen)					+	+	+					
Brillia modesta (Meigen)									+			
Cricotopus annulator Goethebuer							+	+				
C. bicinctus (Meigen)					+		+	+	+	+		
C. festivellus (Kieffer)					+							
C. flavocinctus (Kieffer)									+			
C. pallides (Edwards)						+	+	+	+			
C. similis Goethebuer						+	+	+	+			
C. tremulus (L.)							+					
C. trifascia Edwards									+			
C. albiforceps (Kieffer)					+			+				
C. (Isocladius) pilitarsus (Zett)									+			
C. (Isocladius) (sylvestris?)								+	+			
C. (Isocladius) tricinctus (Meigen)							+					
Eukiefferiella brevicalcar (Kieffer)							+	+	+			
E. claripennis (Lundbeck)									+			
E. clypeata (Kieffer)					+	+	+		+			
E. coerulescens (Kieffer)									+			
E. devonica (Edwards)	+							+	+	+	+	+
E. discoloripes Goethebuer						+		+	+	+		
E. ikleyensis (Edwards)						+	+	+	+	+		
E. verralli (Edwards)							+	+	+	+	+	
Heterotrissocladius marcidus (Walker)							+					
Paratrichocladius skirwithensis (Edwards)						+	+	+	+	+		
Prodiamesa olivacea (Meigen)									+			
Psectrocladius psilopterus (Kieffer)							+		+			
Synorthocladius semivirens (Kieffer)		+					+	+	+			

Appendix 4 (continued).

	J	F	M	A	M	J	J	A	S	O	N	D
Orthocladius sp. 'A' Pinder				+	+		+	+	+			
Rheocricotopus dispar (Goethebuer)				+		+						+
Rheocricotopus indet.							+	+				
Bryophaenocladius ictericus (Meigen)										+	+	
Chaetocladius (perennis?)			+									
Chaetocladius (piger?)											+	
Corynoneura lacustris Edwards					+		+	+	+			
Limnophyes (exiguus?)										+		
Metriocnemus (atratulus?)					+							
Smittia pratorum (Goethebuer)					+		+	+	+	+		
Thienemanniella clavicornis (Kieffer)									+	+		
T. vittata (Edwards)				+	+	+	+	+	+			
T. flavescens (Edwards)			+				+	+				
Endochironomus (dispar?)							+	+				
Kiefferulus tendipediformis (Goethebuer)							+					
Microtendipes chloris (Meigen)							+	+	+	+		
Microtendipes (hyalensis?)				+			+					
Nilothauma brayi (Goethebuer)							+	+				
Paratendipes albimanus (Meigen)							+					
Pentapedilum tritum (Walker)						+						
Polypedilum convictum (Walker)								+				
P. cultellatum (Goethebuer)							+		+			
P. quadriguttatum (Kieffer)							+	+	+			
Stictochironomus sticticus (Fabricius)							+					
S. pictulus (Meigen)							+					
Cladotanytarsus vandewulpi (Edwards)							+	+	+			
Paratanytarsus inopertus (Walker)							+	+	+	+		
Tanytarsus arduennensis Goethebuer							+					
T. brundini Lindberg							+		+	+		
T. eminulus (Walker)							+	+	+			
T. gregarius Kieffer							+					
T. heusdensis (Goethebuer)				+			+					
T. signatus (Wulp)								+				
T. triangularis Goethebuer							+	+	+			
Rheotanytarsus distinctissimus Br.							+	+	+			
R. (muscicola?)							+		+			
R. photophilus (Goethebuer)					+	+	+	+	+			
Stempellina bausei (Kieffer)					+			+				

Diptera (Simuliidae)

	J	F	M	A	M	J	J	A	S	O	N	D
Eusimulium aureum Fries									+	+	+	
Simulium reptans (L.)							+		+			
S. monticola Friedericks										+		
S. variegatum Meigen				+			+			+		

150

Appendix 5   Species seen on the Wye Survey, March–July 1977.

Great northern diver	1	Red-legged partridge	3
(*Gavia immer*)		(*Alectoris rufa*)	
Great-crested grebe	A 5	Partridge	3
(*Podiceps cristatus*)		(*Perdix perdix*)	
Little Grebe	B 4	Pheasant	5
(*Tachybaptus ruficollis*)		(*Phasianus colchicus*)	
Cormorant	B 2	Water rail	1
(*Phalacrocorax carbo*)		(*Rallus aquaticus*)	
Grey Heron	B 5	Moorhen	A 5
(*Ardea cinerea*)		(*Gallinula chloropus*)	
Mallard	A 5	Coot	A 5
(*Anas platyrhynchos*)		(*Fulica atra*)	
Teal	B 2	Lapwing	C 5
(*Anas crecca*)		(*Vanellus vanellus*)	
Gadwall	1	Ringed plover	1
(*Anas strepera*)		(*Charadrius hiaticula*)	
Wigeon	1	Little-ringed plover	C 4
(*Anas penelope*)		(*Charadrius dubius*)	
Tufted Duck	B 4	Golden plover	C 5
(*Aythya fuligula*)		(*Pluvialis apricaria*)	
Pochard	B 4	Snipe	C 2
(*Aythya ferina*)		(*Gallinago gallinago*)	
Goldeneye	B 1	Woodcock	D 3
(*Bucephala clangula*)		(*Scolopax rusticola*)	
Red-breasted merganser	A 2	Curlew	C 5
(*Mergus serrator*)		(*Numenius arquata*)	
Goosander	A 5	Bar-tailed godwit	1
(*Mergus merganser*)		(*Limosa lapponica*)	
Greylag goose	1	Green sandpiper	B 2
(*Anser albifrons*)		(*Tringa ochropus*)	
Canada goose	B 4	Wood sandpiper	1
(*Branta canadensis*)		(*Tringa glareola*)	
Mute swan	A 5	Common sandpiper	A 5
(*Cygnus olor*)		(*Tringa hypoleucos*)	
Buzzard	D 5	Redshank	C 5
(*Buteo buteo*)		(*Tringa totanus*)	
Sparrow-hawk	D 3	Greenshank	B 2
(*Accipiter nisus*)		(*Tringa nebularia*)	
Hen Harrier	1	Great black-backed gull	B 1
(*Circus cyaneus*)		(*Larus marinus*)	
Osprey	1	Lesser black-backed gull	B 2
(*Pandion haliaetus*)		(*Larus fuscus graellsii*)	
Hobby	1	Herring gull	B 2
(*Falco subbuteo*)		(*Larus argentatus*)	
Peregrine	3	Common gull	B 1
(*Falco peregrinus*)		(*Larus canus*)	
Merlin	1	Black-headed gull	B 4
(*Falco columbarius*)		(*Larus ribidundus*)	
Kestrel	D 5	Stock dove	5
(*Falco tinnunculus*)		(*Columba oenas*)	

Wood pigeon	5	Marsh tit	D 5
(*Columba palumbus*)		(*Parus palustris*)	
Turtle dove	3	Willow tit	D 5
(*Streptopelia turtur*)		(*Parus montanus*)	
Collared dove	3	Long-tailed tit	5
(*Streptopelia decaocto*)		(*Aegithalos caudatus*)	
Cuckoo	5	Nuthatch	D 5
(*Cuculus canorus*)		(*Sitta europaea*)	
Barn owl	3	Tree-creeper	D 5
(*Tyto alba*)		(*Certhia familiaris*)	
Little owl	D 5	Wren	5
(*Athene noctua*)		(*Troglodytes troglodytes*)	
Tawny owl	D 5	Dipper	A 5
(*Strix aluco*)		(*Cinclus cinclus*)	
Swift	5	Mistle Thrush	5
(*Apus apus*)		(*Turdus viscivorus*)	
Kingfisher	A 5	Fieldfare	2
(*Alcedo atthis*)		(*Turdus pilaris*)	
Green woodpecker	D 3	Song thrush	5
(*Picus viridis*)		(*Turdus philomelos*)	
Great-spotted woodpecker	D 5	Redwing	2
(*Dendrocopus major*)		(*Turdus iliacus*)	
Lesser-spotted woodpecker	D 3	Ring ouzel	1
(*Dendrocopus minor*)		(*Turdus torquatos*)	
Skylark	3	Blackbird	5
(*Alauda arvensis*)		(*Turdus merula*)	
Swallow	5	Wheat-ear	3
(*Hirundo rustica*)		(*Oenanthe oenanthe*)	
House martin	5	Stonechat	3
(*Delichon urbica*)		(*Saxicola torquata*)	
Sand martin	A 5	Whinchat	D 5
(*Riparia riparia*)		(*Saxicola rubetra*)	
Raven	3	Red-start	D 5
(*Corvus corax*)		(*Phoenicurus phoenicurus*)	
Carrion crow	5	Nightingale	1
(*Corvus corone*)		(*Luscinia megarhynchos*)	
Rook	5	Robin	5
(*Corvus frugilegus*)		(*Erithacus rubecula*)	
Jackdaw	5	Grasshopper warbler	1
(*Corvus monedula*)		(*Locustella naevia*)	
Magpie	5	Reed warbler	1
(*Pica pica*)		(*Acrocephalus scirpaceus*)	
Jay	3	Sedge warbler	A 5
(*Garrulus glandarius*)		(*Acrocephalus schoenobaenus*)	
Great tit	5	Blackcap	D 5
(*Parus major*)		(*Sylvia atricapilla*)	
Blue tit	5	Garden warbler	D 5
(*Parus caeruleus*)		(*Sylvia borin*)	
Coal tit	5	Whitethroat	D 5
(*Parus ater*)		(*Sylvia communis*)	

Lesser whitethroat	D 5	Starling	5
(*Sylvia curruca*)		(*Sturnus vulgaris*)	
Willow warbler	5	Greenfinch	5
(*Phylloscopus trochilus*)		(*Carduelis chloris*)	
Chif-chaff	5	Goldfinch	5
(*Phylloscopus collybita*)		(*Carduelis carduelis*)	
Wood warbler	5	Siskin	1
(*Phylloscopus sibilatrix*)		(*Carduelis spinus*)	
Goldcrest	3	Linnet	5
(*Regulus regulus*)		(*Acanthis cannabina*)	
Spotted flycatcher	D 5	Redpoll	5
(*Muscicapa striata*)		(*Acanthis flammea*)	
Pied flycatcher	D 5	Bull-finch	5
(*Ficedula hypoleuca*)		(*Pyrrhula pyrrhula*)	
Dunnock	5	Crossbill	3
(*Prunella modularis*)		(*Loxia curvirostra*)	
Meadow pipit	5	Chaffinch	5
(*Anthus pratensis*)		(*Fringilla coelebs*)	
Tree pipit	3	Brambling	1
(*Anthus trivialis*)		(*Fringilla montifringilla*)	
Pied wagtail	A 5	Corn bunting	1
(*Motacilla alba yarrelli*)		(*Emberiza calandra*)	
White wagtail	1	Yellow-hammer	3
(*Motacilla alba alba*)		(*Emberiza citrinella*)	
Grey wagtail	A 5	Reed bunting	A 5
(*Motacilla cinerea*)		(*Emberiza schoeniclus*)	
Yellow wagtail	A 5	House sparrow	5
(*Motacilla flava flavissima*)		(*Passer domesticus*)	
Blue-headed wagtail	1	Tree sparrow	5
(*Motacilla flava flava*)		(*Passer montanus*)	

Notes: A, B, C, D (see p. 105).

(see p. 105)

1 – Species which occurred as scarce visitors.
2 – Species which occurred as regular visitors.
3 – Species which were probably breeding in the area.
4 – Species which bred on nearby waters.
5 – Species which were definitely breeding.

Bibliography

Anon, 1906. Guide to Llandrindod Wells. Oswestry and Wrexham: Woodall, Minshall, Thomas.

Armitage, E., 1914. Vegetation of the Wye Gorge at Symond's Yat. J. Ecol. 2: 98–109.

Armitage, P. D., 1977. Development of the macro-invertebrate fauna of Cow Green Reservoir (upper Teesdale) in the first five years of its existence. Freshwater Biol. 7: 441–454.

Armitage, P. D., 1978. Downstream changes in the composition, numbers and biomass of bottom fauna in the Tees below Cow Green Reservoir and in an unregulated tributary, Maize Beck, in the first five years after impoundment. Hydrobiologia 58: 145–156.

Armitage, P. D. and Capper, M. H., 1976. The numbers, biomass and transport downstream of micro-crustaceans and *Hydra* from Cow Green Reservoir (upper Teesdale). Freshwater Biol. 6: 425–432.

Badcock, R. M., 1949. Studies on stream life in tributaries of the Welsh Dee. J. Anim. Ecol. 18: 193–208.

Bass, J. A. B., 1976. Studies on *Ephemerella ignita* (Poda) in a chalk stream in Southern England. Hydrobiologia 49: 117–121.

Birtles, A. B., 1977. Computer models of conservative water quality determinands in the River Severn. Proc. Int. Assoc. Water Pollut. Res. Spec. Conf. on River Basin Management, Essen., Sept., 1977, Paper No. 20.

Birtles, A. B. and Brown, S. R. A., 1978. Computer prediction of the changes in river water quality regimes following large scale inter-basin transfers. Int. Symp. on Modelling the Water Quality of the Hydrological Cycle, Baden, Sept., 1978.

Boon, P. J., 1978. The pre-impoundment distribution of certain Trichoptera larvae in the North Tyne river system (Northern England). with particular reference to current speed. Hydrobiologia 57: 167–174.

Boon, P. J., 1979. Studies of the spatial and temporal distribution of larval Hydropsychidae in the North Tyne river system (Northern England) Arch. Hydrobiol. 85: 336–359.

Bongourd, S. M. and Parker, J. S., 1975. The B-chromosome system of *Allium schoenopraesum.* Chromosoma 53: 273–282.

Brooker, M. P. and Hemsworth, R. J., 1978. The effect of an artificial discharge of water on invertebrate drift in the R. Wye, Wales. Hydrobiologia 59: 155–163.

Brooker, M. P. and Morris, D. L., 1978. The production of two species of Ephemeroptera (*Ephemerella ignita* Poda and *Rhithrogena semicolorata* Curtis) in the upper reaches of the R. Wye. Verh. int. verein. theor. angew. Limnol. 20: 2600–2604.

Brooker, M. P. and Morris, D. L., 1980a. A survey of the macroinvertebrate riffle fauna of the River Wye. Freshwater Biol. 10: 437–458.

Brooker, M. P. and Morris, D. L., 1980b. *Potamanthus luteus* (Linnaeus) (Ephemeroptera: Potamanthidae) in the River Wye. Entomol. Gaz. 31: 247–251.

Brooker, M. P., Morris, D. L. and Hemsworth, R. J., 1977. Mass mortalities of adult salmon (*Salmo salar*) in the R. Wye, 1976. J. Appl. Ecol. 14: 409–417.

Brooker, M. P., Morris, D. L. and Wilson, C. J., 1978. Plant–flow relationships in the R. Wye catchment. EWRS 5th Int. Symp. Aquatic Weeds, 5–8 Sept. 1978, 63–70.

Brown, E. H., 1960. The relief and drainage of Wales. Cardiff: University Press.

155

Bryan, G. H., 1894. The valley of the Wye. Sci. Gossip 1: 202–203.

Centre for Agricultural Strategy, 1978. The future of upland Britain. Conference. Ed. E. B. Tranter. CAS paper 2, University of Reading.

Church, B. M., 1976. Use of fertilisers in England and Wales, 1976. Rep. Rothampstead Exp. Sta. 2: 189–193.

Claridge, P. N. and Gardner, D. C., 1978. Growth and movements of the twaite shad, *Alosa fallax* (Lacepede), in the Severn Estuary. J. Fish Biol. 12: 203–211.

Cragg, B., Fry, J. C., Bacchus, Z. and Thurley, S. S., 1980. The aquatic vegetation of Llangorse Lake, Wales. Aquat. Bot. 8: 187–196.

Cragg-Hine, D. and Jones, J. W., 1969. The growth of dace *Leuciscus leuciscus* (L.), roach *Rutilus rutilus* (L.) and chub (*Squalius cephalus* (L.)) in Willow Brook, Northamptonshire. J. Fish. Biol. 1: 59–82.

Crisp, D. T., 1966. Input and output of minerals for an area of Pennine moorland. J. Appl. Ecol. 3: 327–348.

Crisp, D. T., 1977. Some physical and chemical effects of Cow Green (uper Teesdale) impoundment. Freshwater Biol. 1: 109–120.

Crisp, D. T., 1981. A desk study of the relationship between temperature and hatching time of five species of salmonid fishes . Freshwater Biol. 11: 361–368.

Crisp, D. T., Mann, R. H. K. and McCormack, J. C., 1974. The populations of fish at Cow Green, upper Teesdale. before impoundment. J. Appl. Ecol. 11: 969–996.

Dawson, F. H., 1978. Seasonal effects of aquatic plant growth on the flow of water in a small stream. EWRS 5th Int. Symp. Aquatic Weeds, 5–8 Sept. 1978, 71–78.

Day, F., 1890. Notes on the fish and fisheries of the Severn. Proc. Cotswold Nat. Field Club 9: 202–219.

Earp, J. R. and Hains, B. A., 1971. British Regional Geology. The Welsh Borderland. HMSO, 118 pp.

Edwards, A. M. C., 1973a. The variation of dissolved constituents with discharge in some Norfolk rivers. J. Hydrol. 18: 219–242.

Edwards, A. M. C., 1973b. Dissolved load and tentative solute budgets of some Norfolk catchments. J. Hydrol. 18: 201–217.

Edwards, R. W. (in press). Predicting the environmental impact of a major reservoir development. Symp. Br. Ecol. Soc. on Environmental Impact Assessment, Essex, 1980.

Edwards, R. W., Hughes, B. D. and Read, M. W., 1975. Biological survey in the detection and assessment of pollution. In: The Ecology of resource degradation and renewal, Ed. M. J. Chadwick and G. T. Goodman. 15th Symp. Br. Ecol. Soc., 139–156.

Edwards, R. W., Oborne, A. C., Brooker, M. P. and Sambrook, H. T., 1978. The behaviour and budgets of selected ions in the Wye catchment. Verh. int. verein. Theor. angew. Limnol. 20: 1418–1422.

Egglishaw, H. J., 1970. Production of salmon and trout in a stream in Scotland. J. Fish. Biol. 2: 117–136.

Egglishaw, H. J. and Morgan, N. C., 1965. A survey of the bottom fauna of streams in the Scottish Highlands. Part II. The relationship of the fauna to the chemical and geological conditions. Hydrobiologia 26: 173–183.

Elliott, J. M., 1975. The growth rate of brown trout (*Salmo trutta* L.) fed on maximum rations. J. Anim. Ecol. 44: 805–821.

Elliott, J. M., 1978. Effect of temperature on the hatching time of eggs of *Ephemerella ignita* (Poda) (Ephemeroptera: Ephemerellidae) Freshwater Biol. 8: 51–58.

Elliott, J.M. and Corlett, J., 1972. The ecology of Morecambe Bay. IV. Invertebrate drift into and from the R. Leven. J. Appl. Ecol. 9: 195–206.

Ellison, F. B., 1935. Shad. The Woolhope Naturalists Field Club Transactions. 135–139.

Furet, J. E., 1979. Algal studies on the River Wye system. Unpublished Ph.D. Thesis, University of Wales.

Gardner, M. L. G., 1976. A review of factors which may influence the sea-age and maturation of Atlantic salmon, *Salmo salar* L. J. Fish. Biol. 9: 289–327.

Gee, A. S. and Edwards, R. W. (in press). Recreational exploitation of the Atlantic salmon in the River Wye. Int. Symp. on Fishery Resources Allocation, Vichy, 1980, EIFAC.

Gee, A. S. and Milner, N. J., 1980. Analysis of 70-year catch statistics for Atlantic salmon (*Salmo salar*) in the River Wye and implications for management of stocks. J. Appl. Ecol. 17: 41–57.

Gee, A. S., Milner, N. J. and Hemsworth, R. J., 1978a. The effect of density on mortality in juvenile Atlantic salmon (*Salmo salar*). J. Anim. Ecol. 47: 497–505.

Gee, A. S. Milner, N. J. and Hemsworth, R. J., 1978b. The production of juvenile Atlantic salmon, *Salmo salar*, in the upper Wye, Wales. J. Fish. Biol. 13: 439–451.

George, T. N., 1970. British Regional Geology. South Wales. HMSO, 152pp.

Gibbons, D. R. and Salo, E. O., 1973. An annotated bibliography of the effects of logging on fish of the Western United States and Canada. USDA. Foreign Serv. Gen. Tech. Rep., PNW-10.

Gilbert, H. A., 1929. The tale of a Wye fisherman. London: Methuen.

Gissing, T. W., 1853. Notes of a biological excursion down the Wye. Phytologist 4: 1053–1055.

Hansford, R. G. and Ladle, M., 1979. The medical importance and behaviour of *Simulium austeri* Edwards (Diptera: Simuliidae) in England. Bull. Entomol Res. 69: 33–41.

Harper, P. P., 1978. Variations in the production of emerging insects from a Quebec stream. Verh. int. verein. theor. angew Limnol. 20: 1317–1323.

Harriman, R. and Morrison, B. R. S., 1981. Forestry, fisheries and acid rain in Scotland. Scottish Forestry. 35: 89–95.

Haslam, S. M., 1978. River plants: the macrophytic vegetation of watercourses. Cambridge University Press.

Hellawell, J. M., 1969. Age determination and growth of the grayling *Thymallus thymallus* (L.) of the River Lugg, Herefordshire, tributary of the River Wye. J. Fish. Biol. 1: 373–382.

Hellawell, J. M., 1971a. The autecology of the chub, *Squalius cephalus* (L.) of the River Lugg and Afon Llynfi. I: Age determination, population structure and growth. Freshwater Biol. 1: 29–60.

Hellawell, J. M., 1971b. The autecology of the chub, *Squalius cephalus* (L.), of the River Lugg and Afon Llynfi. II: Reproduction. Freshwater Biol. 1: 135–148.

Hellawell, J. M., 1971c. The autecology of the chub, *Squalius cephalus* (L.), of the River Lugg and Afon Llynfi. III: Diet and feeding habits. Freshwater Biol. 1: 369–387.

Hellawell, J. M., 1972. The growth, reproduction and food of the roach, *Rutilus rutilus* (L.) of the River Lugg, Herefordshire. J. Fish.Biol. 4: 469–486.

Hellawell, J. M., 1973. Fish distribution in the R. Wye. J. Inst. Fish. Mgmt. 4: 19–20.

Hellawell, J. M., 1974. The ecology of populations of dace *Leuciscus leuciscus* L. from two tributaries of the River Wye, Herefordshire, England. Freshwater Biol. 4: 577–604.

Hemsworth, R. J., 1979. Invertebrate drift in the upper Wye catchment. Unpublished M.Sc. Thesis, University of Wales.

Hemsworth, R. J. and Brooker, M. P., 1979. The rate of downstream displacement of macroinvertebrates in the upper Wye, Wales. Holarctic Ecol. 2: 130–136.

Hildrew, A. G. and Edington, J. M., 1979. Factors facilitating the coexistence of hydropsychid caddis larvae (Trichoptera) in the same river system. J. Anim. Ecol. 48: 557–576.

Hopper, F. N., 1978. Faunal studies on the River Elan. Unpublished M.Sc. Thesis, University of Wales.

Houston, J. A. and Brooker, M. P., 1981. A comparison of nutrient sources and behaviour in two lowland subcatchments of the River Wye. Water Res. 15: 49–57.

Hughes, B. D. and Edwards, R. W., 1977. Flows of sodium, potassium, magnesium and calcium in the R. Cynon, S. Wales. Water Res. 11: 563–566.

Hutton, J. A., 1923. Something about grayling scales. Salm. Trout Mag., Jan., 3–8.

Hutton, J. A., 1930. Rod-fishing for salmon on the Wye. Fish. Gazette (London).

Jones, O. T., 1946. The complex intrusion of Welfield, near Builth Wells, Radnorshire. Quart, J. Geol. Soc. 102: 157–188.

Jones, O. T., 1948. The form and distribution of dolerite masses in the Builth-Llandrindod inlier, Radnorshire. Quart. J. Geol. Soc. 104: 71–98.

Jones, O. T., 1949. An early Ordovician shoreline in Radnorshire. near Builth Wells. Quart. J. Geol. Soc. 105: 65–99.

Jones, T., 1805. History of Brecknockshire, Vol. 1. William and George North, Brecon.

Jones, R. and Benson-Evans, K., 1974. Nutrient and phytoplankton studies of Llangorse Lake. Field Stud. 4: 61–65.

Judd. W. W., 1953. A study of the population of insects emerging as adults from the Dundas Marsh, Hamilton, Ontario, 1948. Am. Midl. Nat. 49: 810–824.

Kipling, C. and Frost, W. E., 1970. A study of the mortality, population numbers, year-class strengths, production and food consumption of pike. *Esox lucius* (L.) in Windermere from 1944 to 1962. J. Anim. Ecol. 39: 115–157.

Lack, T. J., 1971. Quantitative studies on the phytoplankton of the Rivers Thames and Kennet at Reading. Freshwater Biol. 1: 213–224.

Langbein, W. B. and Schumm, S. A., 1958. Yield of sediment in relation to mean annual precipitation. Am. Geophys. Union Trans. 39: 1076–1084.

Lavis, M. E. and Smith, K., 1972. Reservoir storage and the thermal régimes of rivers, with special reference to the River Lune, Yorkshire. Sci. Total Environ. 1: 81–90.

Le Cren, E. D., 1973. The population dynamics of young trout (*Salmo trutta*) in relation to density and territorial behaviour. Rapp. PV. Reun. Cons. Int. Explor. Mer 164: 241–246.

Lilley, A. J., Brooker, M. P. and Edwards, R. W., 1979. The distribution of the crayfish, *Austropotamobius pallipes* (Lereboullet). in the upper Wye catchment, Wales. Nat. Wales 16: 195–200.

Maitland, P. S., 1964. Quantitative studies on the invertebrate fauna of sandy and stony substrates in the River Endrick, Scotland. Proc. Roy. Soc. Edinburgh Sect B 68: 277–301.

Maitland, P. S., 1965. The distribution, life-cycle and predators of Ephemerella ignita (Poda) in the R. Endrick, Scotland. Oikos 16: 48–57.

Maitland, P. S., 1966. The fauna of the R. Endrick. Studies on Loch Lomond 2. Glasgow: Blackie, 194pp.

Maitland, P. S ., 1972. Key to British freshwater fishes. Sci. Publ. Freshwater Biol. Assoc. 27: 139pp.

Mann, R. H. K., 1971. The populations, growth and production of fish in four small streams in Southern England. J. Anim. Ecol. 40: 155–190.

Mann, R. H. K., 1974. Observations on the age, growth, reproduction and food of dace, *Leuciscus leuciscus* (L.) in two rivers in Southern England. J. Fish. Biol. 6: 237–253.

Mann, R. H. K., 1976. Observations on the age, growth, reproduction and food of chub, *Squalius cephalus* (L.) in the River Stour, Dorset. J. Fish. Biol. 8: 265–288.

Merry, D. G. and Slater, F. M., 1978. Plant colonisation under abnormally dry conditions of some reservoir margins in Mid-Wales. Aquat. Bot. 5: 149–162.

Miller, A. A., 1935. The entrenched meanders of the Herefordshire Wye. Geogr. J. 85: 160–178.

Milner, N. J., Gee, A. S. and Hemsworth, R. J., 1978. The production of brown trout, *Salmo trutta*, in tributaries of the upper Wye, Wales. J. Fish. Biol. 13: 599–612.

Milner, N. J., Gee, A. S. and Hemsworth, R. J., 1979. Recruitment and turnover of populations of brown trout, *Salmo trutta*, in the upper River Wye, Wales. J. Fish. Biol. 15: 211–222.

Monnington, H. W., 1889. Notes on the flora of the Wye. Sci. Gossip 25: 67–68.

Morgan, N. C. and Waddell, A. B., 1960. Insect emergence from a small trout loch and its bearing on the food supply of fish. Sci. Invest. Freshwater Fish. Scot. 25: 1–39.

Morris, D. L., 1981. Life cycles and aspects of emergence and flight behaviour of aquatic insects from the upper Wye, Wales. Unpublished M.Sc. Thesis, University of Wales.

Morris, D. L. and Brooker, M. P., 1979. The vertical distribution of macroinvertebrates in the substratum of the upper reaches of the R. Wye, Wales. Freshwater Biol. 9: 573–583.

Muenscher, W. C., 1944. Aquatic Plants of the United States. New York: Comstock.

Murchison, R. I., 1839. The Silurian system. London: Murray.

Neill, R. M., 1938. The food and feeding of brown trout (*Salmo trutta* L) in relation to the organic environment. Trans. Roy. Soc. Edinburgh 59: 481–520.

Newson, M. D., 1976. The physiography, deposits and vegetation of the Plynlimon catchments. Institute of Hydrology, Unpublished report no. 30, 44pp.

Newson, M. D., 1979. The results of ten years' experimental study on Plynlimon, Mid-Wales and their importance for the water industry. J. Inst. Eng. Sci. 33: 321-333.

Oborne, A. C., 1981. The application of a water-quality model to the River Wye, Wales. J. Hydrol. 52: 59-70.

Oborne, A. C., Brooker, M. P. and Edwards, R. W., 1980. The chemistry of the River Wye. J. Hydrol. 45: 233-252.

Orloci, L., 1975. Multivariate analysis in vegetation research. The Hague: Junk.

Owens, M., Garland, J. H. N., Hart, I. C. and Wood, G., 1972. Nutrient budgets in rivers. Symp. Zool. Soc. London 29: 21-40.

Page, W., Ed., 1908. The Victoria History of the County of Hereford. London: Constable.

Paull, L. M., 1978. Benthic invertebrates and deposits in the River Elan. Unpublished M.Sc. Thesis, University of Wales.

Perring, F., 1957. Llandrindod Wells August 4th-11th 1956. Field Meetings. Proc. Bot. Soc. Br. Is. 11: 416-418.

Randerson, P. F., Edwards, R. W. and Shaw, K., 1978. An economic evaluation of the demand for recreational fishing in areas in Wales 203-218. Proc. Conf. on Recreational Freshwater Fisheries. Water Research Centre, Oxford, 1977.

Ratcliffe, D., 1977. A nature conservation review, Vol. 2. Site accounts. Cambridge University Press, 320pp.

Ricker, W. E., 1954. Stock and recruitment. J. Fish. Res. Board. Can. 11: 559-623.

Riddelsdell, H. J., 1910. The botany of the county around Builth Wells. Proc. Cotswold Nat. Field Club 17: 57-61.

Roos, T., 1957. Studies on upstream migration in adult stream dwelling insects. Rep. Inst. Freshwater Res. Drottningholm 38: 167-193.

Round, F. E., 1956. The phytoplankton of three water supply reservoirs in central Wales. Arch. Hydrobiol. 52: 457-469.

Round, F. E., 1957. The benthic algal flora of the three city of Birmingham water works reservoirs in Central Wales. Arch. Hydrobiol. 52: 562-573.

Scullion, J. and Edwards, R. W., 1980. The effect of coal industry pollutants on the macro-invertebrate fauna of a small river in the South Wales coalfield. Freshwater Biol. 10: 141-162.

Sculthorpe, C. D., 1967. The biology of aquatic vascular plants. London: Edward Arnold.

Sharrock, J. T. R., 1976. The atlas of breeding birds in Britain and Ireland. British Trust for Ornithology, Tring.

Shaw, J. S., 1977. The freshwater fish and fisheries of Wales. Unpublished M.Sc. Thesis, University of Wales.

Smith, R. V., 1976. Nutrient budget of the R. Main, Co. Antrim. Tech. Bull. Minist. Agric. Fish. Food, London no. 32: 315-339.

Smyly, W. J. P., 1957. The life-history of the bull-head or Miller's thumb (*Cottus gobio* L.) Proc. Zool. Soc. London 128: 431-450.

Sutcliffe, D. W. and Carrick, T. R., 1973a. Studies on mountain streams in the English Lake District. II. Aspects of water chemistry in the R. Duddon. Freshwater Biol. 3: 543-560.

Sutcliffe, D. W. and Carrick, T. R., 1973b. Studies on mountain streams in the English Lake District I. pH, calcium and the distribution of invertebrates in the River Duddon. Freshwater Biol. 3: 437-462.

Symons, P. E. K., 1979. Estimated escapement of Atlantic salmon (*Salmo salar*) for maximum smolt production in rivers of different productivity. J. Fish. Res. Board Can. 36: 132-140.

Tate-Regan, C., 1911. The freshwater fishes of the British Isles. Methuen.

Thompson, R. W. S., 1954. Stratification and overturn in lakes and reservoirs. J. Inst. Water Eng. 8: 19-52.

Toms, I. P., Wood, G. and Owens, M., 1975. Grafham water: some effects of the impoundment of nutrient-rich water. Pap. Proc. Water Res. Cent. Symp. on The Effects of Storage on Water Quality. University of Reading, March 24-26, 1975, 121-161.

Troake, R. P., Troake, L. E. and Walling, D. E., 1976. Nitrate loads of South Devon streams. Tech. Bull. Minist. Agric. Fish. Food, London, no. 32: 340-351.

Walford, L. A., 1946. A new graphic way of describing the growth of animals. Biol. Bull. Mar. Biol. Lab., Woods Hole 90: 141-147.

Ward, J. V., 1975. Downstream fate of zooplankton from a hypolimnial release mountain reservoir. Verh. int. verein. theor. angew. Limnol. 19: 1798-1804.

Water Resources Board, 1973. Water resources in England and Wales. HMSO.

Watson, A. K., 1979. A general survey of the benthic invertebrates in tributaries of the upper Wye. Unpublished M.Sc. Thesis, University of Wales.

Welsh, F. B. A. and Trotter, F. M., 1960. Geology of the country around Monmouth and Chepstow. Mem. Geol. Surv. HMSO.

Westlake, D. F., 1963. Comparisons of plant productivity. Biol. Rev. 38: 385-425.

Williams, W. P., 1967. The growth and mortality of four species of fish in the River Thames at Reading. J. Anim. Ecol. 36: 695-720.

Williamson, K., 1975. Waterways bird survey report 1974. Bird Stud. 22: 53-54.

Wisniewski, P. J., 1978. The riffle and pool invertebrates of the upper Wye. Unpublished M.Sc. Thesis. University of Wales.

Wright, C. E., 1976. Once in 1000 years? Water 9: 411-423.

Wye River Authority, 1972. Survey of water resources and demands. Wye River Authority, Hereford.

Wye River Division, 1977. The 1976 drought. Wye River Division, Hereford.

Index*

* Numbers in **bold-face type** refer to pages which contain figures or tables.

exploitation   **98**, 99–102, **101**
growth   94, **95**
mortality   95, **96**, 102, **102**
production   94, 96, 97
spawning   94, 95
Sampling sites   **4**, 27, 53, 111
Sewage effluent. *See* Effluent
Shad   82
Silica   **16**, 17, 18, **18**, 45, 49
Silurian   1, 2, 5, 34
Smolts   95
Sodium   18, 20
Spa waters   11, **12**
Spatial distributions
   birds   105–109, **106, 107**
   fish   79–81, **81**
   macroinvertebrates   55–68, **56, 58, 59, 63, 66, 67**
   otters   **104**, 110–111
   plants   28–40, **28, 30, 36, 37, 41**
   water quality   13–15
Spawning. *See* Fish
Spraint   111
Stratification (reservoir)   25
Suspended solids   18

Temperature   12, **19**, 20, 25, 42, 43, **43**
   effects on fish diet   86
   effects on fish growth   87, 94, **95**, 121
   effects on fish mortality   102, **102**
Total dissolved solids   **13**, 15, 21, 54
Total hardness   119
Transfer, water   114
Triassic   1, 3

Trichoptera   62–64, **63**, 72
Trothy, River   11, **17**, 18, 21, 80, 81
Trout   91–94, **92, 93**

Ulcerative Dermal Necrosis   102
Uplands   5, 113
Urban development   21

Vegetation. *See* Plants

Waders   109
Water quality   11–25, 54, **61**
   fish production   93–94, **93**
   flow relationships   15, **16**, 21, 24, **118**
   ground   17
   model   21, 24, **24**
   of Elan Valley Reservoirs   24, **25**
   of Wye   13–15
   silt   114
   spatial distribution   1, 13–15, **13, 14**
   storage   121
   temporal variation   **14**, 15, 17
Waters
   resources   6, 113, 114–115
   treatment-works   15

Yields
   chemical   20–24, **22, 23**
   water, effects of afforestation   113

Zannichellia palustris   49, **50, 51**
Zinc   **13**, 15
Zooplankton   76–78, 128